A CULTURAL HISTORY OF INSECTS

VOLUME 2

A Cultural History of Insects
General Editor: Gene Kritsky

Volume 1
A Cultural History of Insects in Antiquity
Edited by Gene Kritsky

Volume 2
A Cultural History of Insects in the Medieval Age
Edited by Gene Kritsky

Volume 3
A Cultural History of Insects in the Renaissance
Edited by Sherilyn G.F. Smith

Volume 4
A Cultural History of Insects in the Age of Enlightenment
Edited by Matthew S. Lehnert

Volume 5
A Cultural History of Insects in the Age of Industry
Edited by Carol M. Anelli and Susan W. Fisher

Volume 6
A Cultural History of Insects in the Modern Age
Edited by Robert K.D. Peterson

A CULTURAL HISTORY OF INSECTS

IN THE MEDIEVAL AGE

VOLUME 2

Edited by Gene Kritsky

BLOOMSBURY ACADEMIC
LONDON • NEW YORK • OXFORD • NEW DELHI • SYDNEY

BLOOMSBURY ACADEMIC
Bloomsbury Publishing Plc
50 Bedford Square, London, WC1B 3DP, UK
1385 Broadway, New York, NY 10018, USA
29 Earlsfort Terrace, Dublin 2, Ireland

First published in Great Britain 2024

Series design by Raven Design
Cover image: Honey Illustration from Ibn Butlan's Taqwim al-sihhah or "Maintenance of Health" (Baghdad, eleventh century) © Pictures from History/Bridgeman Images

A catalogue record for this book is available from the British Library.

Library of Congress Cataloging-in-Publication Data

LCCN 2023029669

ISBN: HB: 978-1-3500-0306-4
Pack: 978-1-3500-0321-7
ePDF: 978-1-3502-6652-0
eBook: 978-1-3502-6657-5

Series: The Cultural Histories Series

Typeset by Integra Software Services Pvt. Ltd.
Printed and bound in Great Britain

CONTENTS

ILLUSTRATIONS

FIGURES

TABLE

CONTRIBUTORS

May R. Berenbaum is Professor of Entomology at the University of Illinois at Urbana-Champaign, USA

Silvana Chiesa is Research Fellow in the Department of Food and Drug at the University of Parma, Italy

Julio Ferrer is a Volunteer in the Department of Entomology at the Swedish Museum of Natural History, Sweden

Gene Kritsky is Dean and Professor in the School of Behavioral and Natural Sciences at Mount St. Joseph University, USA

Víctor J. Monserrat is retired from the Departamento de Biodiversidad, Ecología y Evolución, Unidad Docente de Zoología, Facultad de Biología, Universidad Complutense, Madrid, Spain

Rachel Murray is Lecturer in English in the Department of Humanities at Northumbria University, Newcastle upon Tyne, UK

Vazrick Nazari is a Postdoctoral Fellow, Department of Biology University at Padua, Italy

Eileen Reilly is Postdoctoral Research Fellow in the School of Archaeology at University College Dublin, Ireland

Giovanni Sogari is Research Fellow in the Department of Food and Drug at the University of Parma, Italy

SERIES PREFACE

This six-volume set, *The Cultural History of Insects*, is a unique examination of how insects have infested the human experience. The set is divided into six time periods: Antiquity, the Medieval Age, the Renaissance, the Age of Enlightenment, the Age of Industry, and the Modern Age. Within each time period, eight topics are addressed: the state of entomological knowledge during the period; insects and disease; insects and food; insect products; insects' role in mythology and religion; insects as symbols; the portrayal of insects in literature and language; and the representation of insects in art. This structure is meant to allow readers to tailor their reading to their specific interests: for example, if insects' influence in the history of art is of primary interest, the reader can select the art chapter from each volume; if a reader wants to understand the impact of insects on a particular period in human history, then reading the volume on that time period would be appropriate.

The coverage of the topics presented is not encyclopedic. Presented are histories and case studies describing how insects influenced human culture and survival. The geographic focus is generally the Western Hemisphere, but details are included of geographic intersections in which cross-cultural influence plays an important role in various topics and time periods.

Insects were here before humans. They have influenced our evolution, eaten our food, provided us with raw materials, changed our history, and been the muses for our creativity. These six volumes illustrate how insects have helped to shape our culture over the past five thousand years.

Gene Kritsky,
Mount St. Joseph University,
Cincinnati, OH

Introduction

VAZRICK NAZARI

There was a time and place in human history when insects were revered as gods, and that was ancient Egypt. The toil of the humble scarab beetle rolling a ball of dung had a divine significance for the ancient Egyptians: it represented the heavenly cycle, the sun god *Re* rolling across the sky every day and giving life and nourishment to the world, much like the scarab provided food for its young living inside the ball of dung. *Khepri*, the god of the rising sun and creation and renewal of life, was often depicted as a man with a scarab beetle for his head. By the Early Middle Kingdom (2055 BCE), scarabs had become enormously popular in Egypt and, despite the political unrest during this period in Egyptian history, their popularity continued through the New Kingdom and until the end of the Pharaonic era. Their deep symbolic meaning later influenced Christianity, and early Christian saints adopted the symbolism of the scarab, associating it with Christ (Werness 2006).

Much like the Middle Kingdom of Egypt, the Middle Ages in Europe—which stretched for about a thousand years, from the fall of the western Roman Empire in 475 (deposition of Emperor Romulus Augustus by the Barbarian general Flavius Odoacer) to the discovery of America by Christopher Columbus in 1492—were, for the most part, chaotic times. Europe was mired in wars, unrest, disease, and death. Poverty and famine, following deadly epidemics, and locust outbreaks that plagued Europe after the fall of the Roman Empire resulted in an overall decline in literacy that interrupted the pursuit and transmission of scientific knowledge in early medieval Europe. Preservation of classical science and literature, such as the works of the fourth-century BCE Greek philosopher Aristotle and other thinkers of antiquity largely fell on clerics and scholarly theologians.

But this was also the age of new discoveries. Emissaries, merchants and travelers to new lands returned with stories of all kinds of fantastical creatures. In antiquity, stories by authors such as Ctesias of Cnidus, a Greek physician in the court of Artaxerxes II of Persia (fifth century BCE), Deimachus, the Seleucid ambassador in the Indian Court (third century BCE), and the Greek historian Megasthenes (350–290 BCE) introduced new animals and monstrous beings from exotic lands into the early European imagination. These fantastical tales influenced the works of early Greek philosophers such as Aristotle and Pliny, and inspired the authors of the books of the Bible as well as other sacred texts of Christianity. In the Middle Ages, stories by travelers such as Benjamin of Tudela (1130–1173), William of Rubruck (1220–1293), and Marco Polo (1254–1325) followed along the same lines: medieval bestiaries, the predecessors of modern natural history books, teemed with monsters and chimaeras and fantastical beings, all of whom had an equal reality as actual animals. Harpies, basilisks, faucets, and other monstrous beings, influences from ancient Mediterranean, Mesopotamian, Assyrian, Persian, or Hindu cultures, entered European mythology and dominated medieval art and architecture for a long time.

Entomology as a science remained rudimentary in medieval Europe. Despite containing wildly incorrect assertions about animals, Aristotle's biological works, such as *History of Animals* (350 BCE), were among the most influential sources of zoological knowledge throughout the Middle Ages. His now-obsolete theory of spontaneous generation (of living beings emerging from inanimate matter such as flesh, wood, or earthy substances) took firm hold as scientific dogma and continued to be taught in Europe until the mid-seventeenth century (Kern 2005). Aristotle maintained that worms originated in woods or rotting meat, caterpillars in cabbage, and moths in clothes. Bees were thought to have virginal reproduction, fetching their young from flowers; and spiders and wasps were thought to take nourishment from the air (Rendell 1928). It wasn't until 1668 when Italian biologist Francesco Redi was able to demonstrate that flies had to lay eggs on meat before maggots would develop, and Louis Pasteur's classic experiments in the early 1860s finally debunked Aristotle's theory.

Insects as we know them today were not well-defined. Until Linnaeus segregated them in 1758 based on shared morphological characters, insects were all seen as part of a large and loosely defined group of organisms which included earthworms, slugs, snails, some reptiles, and even dragons. In medieval literature, what we know today as insects were placed somewhere between worms (*vermin*) and birds: they were the "creepy-crawlies"; in fact, the word "insect" itself did not enter into English language until 1601, in Holland's translation of Pliny's *Naturalis Historia* (77–79 CE). In this book, Pliny the Elder classified insects not only on the basis of their anatomy, but also based on their advantages or disadvantages for humans. Pliny placed scorpions and spiders together with bees, wasps, hornets, bumblebees, cicadas, beetles, locusts, ants, moths, and mosquitoes (Maraglino 2012). The anonymous second-century text *Physiologus*, a predecessor to medieval bestiaries, was also fraught with biological inaccuracies and mythical animals, usually discussed with a heavy moral overlay (Morge 1973). In this book, ants were given special merit since they represented the human society; honey bees were an inspiration for the collective work of men towards a common good and to obey duly appointed leaders. Dung beetles, with their balls of evil thoughts and heresy, were used as metaphors for the heretic, a stark contrast to their meaning in Egyptian antiquity. Fig wasps, known since the fifth century BCE and discussed in the works of Herodotus, are also mentioned in *Physiologus* as living in the darkness until the fig is broken open (Goor 1965). Some versions of *Physiologus* mentioned dog-sized African ants as well as the "*ant-lion*," a chimaera that is a lion in the front and an ant in the back, signifying the dual nature of the human soul. They represented the internal struggle of human nature due to its dual identity: the lion-faced father ate meat and the ant-faced mother ate grains, so the ant-lion could not consume either; and similarly, Man cannot follow two paths, but only one; i.e., to salvation (Kevan 1992).

Not much survives in writing from the fifth to ninth centuries, otherwise aptly known as the "Dark Ages." Early biological observations during the High and Late Middle Ages were also not free from the old Aristotelian dogma. In the seventh century, archbishop Isidore of Seville in his book *Etymologiae* described nearly three dozen insects, defining "*vermin*" based on their habitat (e.g., vermin of the earth, the air, the flesh, leaves, wood, and clothing). Spiders, slugs, leech, and scorpions accounted among the vermin, but centipedes were included with the reptiles. Ticks, fleas, nits, bedbugs, and louse were among the "vermin of the flesh." Isidore tied the reproduction of bees to carcasses of cattle: in order to create honey bees, the flesh of slain claves must be beaten to create the worms from the rotten gore. Bees sprang from dying flesh of cattle, beetles and hornets from horses, locusts and drones from mules, wasps from asses, and scorpions from crabs (Brehaut 1912). In the ninth century *De rerum Naturis* (On the Nature of Things), Benedictine monk Rabanus Maurus (780–856) also included insects among a multitude of other small animals, vermin, and serpents. But he also tried to make a distinction between theological subject and those related to nature and agriculture, and so the ant-lion lost its mythological status given to it in *Physiologus*, and was presented as what we know today as the myrmeleontid (Schipper 1997).

Insects make rare appearances in Christian artwork predating the fourteenth century, such as tapestries or paintings, but are generally absent from illuminated manuscripts. The illuminated manuscripts produced during the Middle Ages were mostly of religious nature (books of hours, breviaries, pontificals, ordinals, decretals, psalters, oratories, graduals, and other works of devotion), and were richly illustrated with scenes from daily lives of ordinary people, clerics, and royalty. The margins of these manuscripts were often ornamented with elaborate decorative illustrations, also known as "marginalia," incorporating a variety of natural elements such as flowers, birds, and other animals, including insects. One of the oldest of such manuscripts, the Scottish "*Book of Kells*" from 800 CE (Trinity College Dublin, MS 58), includes two small moths hidden within the gothic calligraphy of the Chi-Rho page (Spangenberg 2010). From the late thirteenth century onwards, insects—especially butterflies—begin to appear more frequently in the borders of European manuscripts.

The majority of these exquisite books were commissioned by the nobility and took several years to complete, and often involved several artists. They were prized possessions not only because of their religious content, but also as magnificent works of art. It is therefore worth noting some of the major preoccupations of the nobility in medieval times: hunting and warfare. In the English "*Bird Psalter*" (1280–1290) (Fitzwilliam Museum, MS 2-1954), a bowman is shown aiming at a snipe but also a white *Pieris* butterfly (Nazari 2014). Scenes of hunting of birds and other animals abound in medieval marginalia, as this was a common activity among the nobles and royalty. However, weapons are more often portrayed in the context of warfare rather than hunting, reflecting the violent tenor of the age. Few regions of Europe or Asia remained untroubled by invasion, rebellion or civil war during the thirteenth and fourteenth century, resulting in a gradual change in the way armies were organized and battles conducted. For example, in 1337, just before the outbreak of the war with France, Edward III of England prohibited all sports except archery on punishment of death (Mortimer 2012). On every village green, young men became proficient in the use of longbows, and standards of archery soared. It is not far-fetched therefore to think that for these medieval warriors (as well as hunters) aspiring to improve their skills in archery or swordsmanship, tiny objects moving erratically in the air constituted the ultimate aiming target, and this may have been a routine part of fighter training in the Middle Ages. This practice was probably commonplace and continued even to modern times: Laubin and Laubin (1980) mention that modern American Indian boys practice aiming by shooting arrows at butterflies.

Some of the scenes involving insects in medieval bestiaries and other illuminated manuscripts may be explained by the well-known religious symbolism and iconography that thrived during the Middle Ages (Panofsky 1955). Writers, poets, and philosophers continued to use insects in their works, either as figures of speech, in proverbs and epithets, or to impart moral lessons. Insects were sometimes praised as divine, sometimes condemned as diabolical; some received high attention and valor, others were ignored or deemed insignificant. Some folk notions continued to persist: ladybugs brought good fortune, the blissful cicadas and crickets brought good luck with their music. Scarab beetles in northern Europe (*Tordyvel*, literally "Thor's devil") were thought to bring good luck and were treated with great respect by the Vikings. Similarly, in Scandinavia, jewel beetles (*Cetonia)* were thought to be mythical creatures associated with dwarfs, the jewelry masters and workers of the subterranean mines (Celander 1911). Christianity adopted many of the symbolic attributes of animals from the classical iconography: Jupiter's eagle, the Delphic perception of the snail, and the snake-eating mongoose were all co-opted to represent Christ, while the owl of Minerva became a symbol of the heretic (Ferguson 1959). In medieval Christian literature and artwork, animals were often given one or more didactic moral purpose, sometimes contradictory: unicorns were symbols of purity, while monkeys represented mischief (Walker Vadillo 2013). Snails may have represented humility (Hope 2013) or virginity (Ettlinger 1978). Scorpions and spiders were often seen as demonic, beings associated with temptation, pain, and purgatory, although spiders were sometimes revered for their wisdom and ingenuity. Flies were symbols of the devil, moral malign, or brevity of life on earth; they were associated with death and putrefaction of corpses because of the diet of necrophilous fly larvae (Werness 2006; Taylor 2013). Ants were considered wise, farsighted, and industrious and represented a perfect, organized society and hard collective work; they were sometimes associated with the Christian virtue of prudence (Werness 2006). Bees were associated with virginity (Berenbaum 1995) or structure and hierarchy (Payne 1990); they were better than the ants because aside from social organization they also produced honey and wax. Wasps and dragonflies were symbols of sin, evil, and disorder, while fireflies, mantids, and ichneumons represented Christ in one way or another. Louse were associated with plague and disease, fleas with lust (Brumble 1973), and locusts with famine and devastation associated with the plagues of Egypt (Morge 1973). Crickets were said to like singing so much that it led to their insanity and death (Mermier 1992); similarly, cicadas were noted for always singing and never working. Ant-lions (*myrmekoleon* or *formicaleon*) were associated with the devil, duality of the soul, and hypocrisy (Parker et al. 1844); ladybugs represented the seven sorrows of the Virgin (Yanoviak 2013), and scarab beetles became associated with sinners, dirt, filth, and moral decay. Moths were not revered as much, and some—including the Sphinx moth, *Acherontia atropos*—were considered bad omens; but butterflies for the most part maintained their status as the iconic representations of the soul and resurrection. Within the limited scope of western art, Gagliardi (1976) describes seventy-four different symbolic contexts in which butterflies and moths may appear. It has been argued that the combination of flies, dragonflies, and butterflies in the marginalia of a folio in medieval marginalia is a representation of the death, ascension, and resurrection of Christ (Hassig 1995).

The majority of popular insect-related medieval fables were adopted from *Aesopica*, a collection of tales attributed to the ancient Greek story teller Aesop, and were also full of antiquated misinformation. An Anglo-Saxon poem in the tenth-century "Exeter book" of riddles which talks about moths eating words, alluding to insects as creators and destroyers of the written word, is in fact referring to the book-pest we recognize today as the larval form of the Deathwatch beetle, *Xestobium rufovillosum* (Krapp and Dobbie 1936). Flies, fleas, gnats, lice (and sometimes even butterflies) were also very unpopular. They were often mentioned as insignificant and worthless creatures because of their diminutive size, frailty, ephemerality, and lack of any obvious use. Flies were called "wretched and dreadful little animals" (Gigante 1964). Beetles appear in a few of these satirical fables which, beside a moral message, reflect the fear of the "vermin" infiltrating the human body (Spiegel 1987). Elsewhere in medieval literature there are other examples of insects capable of infiltrating and destabilizing the boundaries of the human body: Chaucer's translation of Boethius' *The Consolation of Philosophy* (524 CE) states that humans have often been killed by the bites of small flies (Benson 2008). "To have a wild worm in one's head" was a medieval figure of speech for madness (Whiting and Whiting 1968). Maggots retained primarily a negative connotation, partly due to the religious concerns about the fate of the body after death: how could resurrection occur in the afterlife if the soul's physical container had been reduced to mush? In "The Ant and the grasshopper," one of the best known medieval fables adopted from *Aesopica*, virtues of hard work and forward planning are espoused, as opposed to the vices of indolence; while in another fable, a beetle punishes a predatory eagle by destroying its eggs, highlighting the indeterminacy of the role of insects in medieval imagination: insects praised as heroes in one tale could be turned into villains in another. In yet another fable, beetles join forces with wasps, hornets, and bees to sting and invade larger beasts, conveying the message of unity in the face of oppression, and upturning of feudal power structures (Spiegel 1987).

Many of the modern-day European superstitious beliefs, folk traditions, and festivals involving insects undoubtedly originated in the Middle Ages, such as the *Entroido* Festival in Laza, Spain, in which the townspeople throw flour, ash, or dirt filled with live biting red ants onto each other or passers-by. The *Tarantella* dance of southern Italy is said to cure the victim of a tarantula bite, but also depression. In folk tradition of many European countries, counting the number of spots on a ladybug's back is said to reveal how many children one is going to have, how much money one will receive, or how many months of good luck are in the horizon. It is believed that if one consumes green vegetables on Maundy Thursday, mosquitoes won't bite them the entire year. In Britain and Ireland it is believed that a buzzing bee at the window means a visitor shall be arriving soon, or that damselflies would use their long bodies to sew your eyelids shut if you sleep next to a stream (Cock-Starkey 2016).

The Hippocratic Corpus of texts dating from the fourth–fifth centuries BCE remained the main source of medical knowledge right into the medieval times. Ideas about the origin and treatment of diseases were not secular, but were heavily based on a worldview in which factors such as destiny, sin, and astrological influences played as great a part as any physical cause. The efficacy of remedies prescribed by medieval doctors depended on the beliefs of the patient and were often subordinate to spiritual interventions: the sick often preferred to place their hopes on the Church and God to heal them from their sickness. Malnutrition, poor hygiene, lack of clean water and basic sanitation, coupled with the proximity of livestock living next to people, facilitated the spread of many diseases transferred by insects, such as stable and horse flies. The causal link between insects and diseases were completely unknown. Major disease vectors, such as flies and mosquitoes, fleas, lice, ticks, and mites, alongside other non-vector biting insects such as bedbugs, were recognized primarily for their nuisance value more than anything else (Aberth 2016). Outbreaks of cholera, tuberculosis, infantile gastroenteritis, diphtheria and smallpox (vectored by houseflies), malaria—a.k.a. "cyclical fever" (by mosquitoes)—typhus, trench fever and relapsing fever in humans and swinepox in pigs (by lice), Justinian/bubonic plagues (by fleas), and sheep scab (by mites) took massive tolls, impacting population growth as well as agriculture and economic growth for many decades (Campbell 2016). Indeed, it wasn't until Antonie van Leeuwenhoek's invention of microscope in 1676 that disease pathogens such as bacteria were first observed; nevertheless, in the Middle Ages no connections were even suggested between, for example, flea bites and the plague. Instead, widespread panic exacerbated by misinformation and superstition resulted in pointing the blame for the epidemics at human scapegoats, such as Jews or lepers, for poisoning people (Hoffmann 2014). However, in the fourteenth and fifteenth centuries the notion that noxious smells spread disease resulted in the construction of more and more private latrines, and inadvertently helped improve the general living conditions in many towns and villages, particularly in the aftermath of the Black Death, where the town dung-hips were moved away from areas of dense human population, taking with them the disease-carrying flies (Mitchell 2015).

Some early medical books from the Middle Ages attempted to provide practical advice on entomological matters, such as apiculture, combating agricultural pests, or household nuisances. The twenty-volume *Geoponika* (tenth century), despite being highly mired in faith and magic, provided often useful advice, such as repelling gnats by applying a mixture of oil, vinegar, and manna to the head and feet, or using fumes from burning herbs (Dalby 2011), although it also included superstitious recommendations dating back to Pliny's *Naturalis Historia*, such as repelling caterpillars by suspending a crayfish or skull of a female horse in the middle of a garden (Smith and Secoy 1975). Use of human waste as manure in the Middle Ages contributed to the spread of intestinal parasites in the food chain and the general ill-health of populations

(Mitchell 2015). Some authors saw lice and intestinal worms as products of the human body, produced when the four "humors" of the body (blood, phlegm, black bile, and yellow bile) were imbalanced (Campbell 2016). Similarly, the principal classification of foods took into account four qualities: hot/cold and dry/wet, criteria set by the second-century physician Galen of Pergamon: all food items were identified based on presence, prevalence, or absence of these qualities, and cautionary advice was given to avoid the presence of insect larvae in stored flour and rice (Capatti and Montanari 2003). Insects were also referred to by their four Galenean qualities: ants were considered hot, and boiling and inhaling steam from whole ant-hills was recommended as a remedy for phlegm. Honey bees, gnats, and wasps were also hot, while bumblebees, houseflies, tree crickets, grasshoppers, and glow-worms were more cold than hot. Honey was considered "very hot" and not recommended for overweight people (Throop 1998).

The western honey bee (*Apis mellifera*) was among the most praised insects since antiquity. Aristotle's account of the many virtues of bees, such as apparent selflessness, relentless industry, social cooperation, and seemingly sexless reproduction, coupled with sweetness and purity of their products (honey, wax, and mead), not only influenced early Christian thinkers, but also led many medieval fabulists to discern parallels between these winged creatures and members of the Christian Church. Medieval perception of the virginal status of bees resulted in symbolic affiliation with the Virgin Mary. Aristotle noted that there are worker bees and drones whose sole job is to protect the king (i.e., the queen); that bees do not tend to attack except when their hive is disturbed; and that the hive is kept meticulously clean (Balme 2014). As such, bees often appeared in medieval poems and allegories of Christian virtue, with their sacramental act of converting one substance to another—flowers to honeycomb, bread to body—rendering them a fitting emblem of transubstantiation (Peck 1991). In the anonymous second-century text *Physiologus*, some attributes (e.g., the death of a honey bee after stinging) were correctly given, but the incorrect gender assignment of the queen ("king") and the drones ("prince") follow the Aristotelian perception. Had the genders been known, the whole religious and spiritual symbolism would have fallen apart.

It is remarkable how a shift in medieval perception of monks and friars, from respect to disdain, is reflected in entomological literature. St. Bernard of Clairvaux (1090–1153) and Thomas Cantipratanus (1233–1248) both associated honey bees with the holy spirit and compared beehives with the orderly life of a monastic community. More than a century later, Chaucer (1387) likened friars to a disorderly swarm of bees destined for hell (Benson 2008). Similarly, the linguistic resemblance of the words fleas, flies, and friars inspired an anonymous poem against the Carmelite brothers in 1475, depicting them as a satanic swarm (Wright and Halliwell 1841).

In Dante's *Divine Comedy* (1320), insects of all descriptions appear in *Inferno* and *Purgatorio*, but only bees grace the pages of *Paradiso*, in a passage that likens the "holy army" of angels to "a swarm of bees, landing upon flowers" (Sisson 2008). In the medieval diet book *De hoensta voluptante et valitudine* (1474), paragraphs are dedicated in praise of bees and discussions on various types of honey. The intriguing association between bees and oratorical skills (sweetness of speech) is probably rooted in the similarity of both words in Hebrew. Poets, saints, and those with gifts for language and speech are said to have their lips touched with honey bees (Winston 2014). For instance, St. Ambrose (340–397), patron saint of beekeepers and protector of bees, tells in *Golden Legend* how bees fly in and out of his mouth as if they were making honey, depositing the honey of theological knowledge on his lips (Vorágine 2014). On the contrary, big black flies (symbols of death) were sometimes depicted in old Gothic paintings flying in the mouth of a criminal sentenced to death (Quentin Matsys, Bruxelles).

Beekeeping in medieval times was a common and profitable occupation, and honey was an important and valuable commodity. Various kinds of beehives existed throughout the Middle Ages (hollow logs, cork, wooden fennel stalks, pottery, and woven out of wicker or straw, a.k.a "skeps"), and surviving images of these hives in illuminated manuscripts show that they had changed very little from antiquity. Hives were often placed in protected spaces in castles, abbeys, and gardens, and were sometimes stacked on top of each other. Many medieval beekeeping practices, such as releasing captured swarms, wearing protective clothing, smoking the hive, etc. are still practiced today, while some others—e.g., killing bees before honey was harvested, "tanging," or beating a small drum to calm the bees down—have become obsolete. A remarkable use of beehives in medieval warfare was to sling beehives into enemy troops as a combat strategy (Herrod-Hempsall 1937).

Due to its many topical uses, as component of beverages or coating for various ailments, honey in the Middle Ages was often regulated, used in international trade, used to pay rent, or demanded as tax by the ruling elite. It was also an important medicinal: a mixture of honey, vinegar, mustard, and pepper was prescribed for liver treatment (Riché 1978). Mead, a.k.a. "honey water," was a popular alcoholic drink (also useful against coughing) made by fermenting the final washings of honey from the comb in a solution with water. Mulsum, a.k.a. "honey wine," was made by adding honey and water to wine (Allsop and Miller 1996). In arts, painters often used honey in tempera paint mixture with water and eggs (sometimes only the yolk) (Encyclopedia of Fine Art 2018). Another product of bees, the beeswax, had many uses in medieval times: in abbeys for liturgies, to produce Easter candles or to make votives; in lost wax casting in metalworking (Hawthorne and Smith 1963); in encaustic painting and as varnish to protect frescoes (Kritsky 2015); as an adhesive (Crocker 2018); or to seal casks, letters, and documents (Watson et al. 2015). Tablets of beeswax were used to write temporary documents (Nelson 1996). Letters were sealed with beeswax mixed with turpentine and vermilion dye, which was heated and dripped on the envelope before the signet seal was stamped on it.

Among other popular products made by insects in medieval times were two different types of dye used by scribes: a deep-black ink called "iron gall," the principal ingredient of which was crushed and fermented oak gall, curious ball-shaped lumps on oak leaves and twigs that are formed when a cynipid wasp (*Biorhiza pallida*) lays its eggs (Clemens and Graham 2007), as well as the scarlet ink known as cochineal red, made from the ground shells of *Dactylopius coccus* scale insects (Eastaugh et al. 2004). The name for the crimson shade "Vermilion" comes from *vermes*, possibly due to its similarity to the red color produced by cochineal insects (Glick et al. 2005). The "Kermes" (crimson) color used in dying silk was also a product of another scale insect on oak, *Kermes vermilio*. Another black ink, known as "lampblack," was made from a mixture of water, soot, gum Arabic, and honey (Meri 2006). Shellac, a resin secreted by the female Indian lac bug (*Kerria laca*, Coccoidea) which has been in use in the Indian subcontinent and Southeast Asia for millennia, arrived in Europe in the thirteenth century. Its use is first reported as an ingredient in artists' paint mixture in Spain, and later as varnish throughout Europe (Merrifield 1849).

By the High Middle Ages, entomological knowledge gradually began to gain foot in Europe, mainly among the religious thinkers who had free time to pursue knowledge. Authors such as the Franciscan monk Bartholomaeus Anglicus (*De Proprietatibus Rerum* [On the Properties of Things], 1240), theologian Thomas Cantipratanus (*Liber de Natura Rerum*, 1233–1248), and bishop Albertus Magnus (*De Animalibus*, 1255–1270) produced books that had high impacts on the zoological knowledge of the subsequent centuries. Magnus restored Aristotle's zoological works into curricula, gave them more credit than works of Pliny the Elder, and argued that Aristotle's views were not against Christian philosophy. But he was also critical of Aristotle's works and in many cases offered his own interpretations and direct observations. Unlike Aristotle or other early natural philosophers, Magnus gave external morphology a central role in his classifications, argued for intelligence and perfection in animals, and paid special attention to species behavior and ecology. He noted arthropod's body segments, number of legs and wings, differentiation of appendage morphology, and internal anatomy. He was the first to associate specific structures in insect anatomy with particular senses (e.g., proboscis with taste). Of the fifty or so species of *vermin* described by him, fewer than three dozen are insects (Morge 1973).

The books by both Bartholomaeus Anglicus and his student Thomas Cantipratanus contained some of the earliest advice given for biological control of insects, such as using Spanish fly (*Lytta vesicatoria,* Cantharidae) to combat fleas (Morge 1973). Spanish flies were known since antiquity for their many properties as aphrodisiacs, ointment to cauterize infected wounds and sores, or to provoke abortions. Due to their high potency and toxicity, cases of poisoning or homicides with Spanish flies were also frequent. Pliny described the death of an ox who accidentally consumed a beetle in the pasture, possibly a cantharid. In Germany, cantharids were considered a sign of arrival of spring, collected and carried in processions, reminiscent of the ancient cults of Dionysus and Bacchus, where cantharids were consumed as an aphrodisiac. These properties seem to have been all but forgotten in the Middle Ages, although they made a comeback during the renaissance (Hedin 1914). In northern Europe, ladybugs were sometimes crushed and applied as anesthetic for toothache (Tillhagen 1958); they do in fact produce an alkaloid (xylocaine), which is an anesthetic.

Other medieval texts had different remedies to rid oneself of fleas or bedbugs: *Geoponika* (tenth century) recommended placing ground oleander leaves in a hole on the floor to attract fleas, or sprinkling the room with a mixture of water, ground wild cumin, and squirting cucumber seeds (Dalby 2011). *Physica* (twelfth century), a nine-volume pharmacological text by Benedictine abbess St. Hildegard of Bingen (1098–1179), recommended heating dirt on an earthenware tile until it is entirely free of moisture and sprinkling it in the bed to eliminate fleas. Bartholomaeus Anglicus recommended "wormwood, the leaves of the wild fig-tree, and colocynth," some of which are still used today as herbal products to repel fleas (Morge 1973). Some of these books had detailed instructions on de-lousing, a common practice in medieval times (Smith 2007). *Geoponika* recommended horse-hair and a soap-vinegar cocktail to fight mosquitoes (Dalby 2011). Controlling animal parasites was also important: records of medieval manors show expenditures on various remedies for sheep scab, a disease vectored by psoroptid mites (Slavin 2016).

Throughout medieval times, in large parts of Europe it was believed that pests were sent by God to punish sinners, and therefore pest control was based on prayer. Holy water was used for centuries as a pesticide, sprinkled in areas invaded by locusts, weevils, caterpillars, and flies; for example, in 1603 in the city of Fremona (in modern-day Ethiopia) holy water was sprinkled to free its residents from invading locusts (Purchas 1614). Other superstitious and bizarre pest control methods included sprinkling weasel ashes or water in which a cat had been bathed over fields to drive away mice; capturing a rodent, castrating it, and releasing it among other rodents to deter them; putting castor oil plants in afflicted fields to drive away moles; and hanging garlic around flock leaders' necks to protect sheep from wolves (Leeson 2013; Vatomsky 2017). The courts in many countries (including France, Italy, and Switzerland) tried animals for crimes against people, property, and God. Domestic animals (such as pigs, dogs, goats, cows, horses, and sheep) fell under the jurisdiction of civil and criminal courts, and if convicted, would have been summarily executed by hanging or burning after a guilty verdict. On the other hand, ecclesiastical tribunals handled all "*vermin*," from rats and moles to serpents, eels, leeches, snails, insects, and worms (caterpillars), since these creatures were, strictly speaking, not under Man's control and thus they were often punished by excommunication from the Catholic Church. In 1479 in Berne, Switzerland, caterpillars (of a species of beetle called an "*inger*") were excommunicated and banished by the archbishop. In 1493 Lausanne, maybugs (*Melolontha* larvae, Coleoptera) were banished for their misbehavior (Hagen 1884). Another lawsuit was brought against flies in Mainz, Germany, in the fifteenth century. Ecclesiastic judges showed impressive fairness toward vermin in the trial, and when the flies refused to appear in court after being justly summoned, the court took pity on their small size and young age and appointed them more adequate defense counsel to prevent further no-shows (Leeson 2013). A set of regulations written by jurist Bartholomaeus Chassanaeus of Burgundy (1517) tells the story of how legal procedures were carried out against grasshoppers:

> A court would be convened upon written request, a Judge appointed, and a prosecuting and defense lawyer assigned. The prosecutor would present the case against the grasshopper and demand that they be found guilty and burned. The defense would argue that the demand was illegal because first the grasshopper had to be requested to leave the country within a specified period of time. If at the expiration of this period they had not left, the proper sentence was excommunication.
>
> (Howard 1895)

Jurists would then argue for the rights of the birds who would suffer from hunger at the extradition of the grasshoppers. In many instances, mitigation of God or the Pope was sought to alleviate the situation. It has been argued that the involvement of the Catholic Church in vermin trials was partly to increase tithe revenue in areas of high tithe evasion, but also due to superstition and lack of practical pest-control knowledge (Leeson 2013). Sometimes communities sued pests collectively: in 1659, the Italian communes of Chiavenna, Mese, Gordona, Prada, and Samolico banded together to prosecute caterpillars they charged with trespassing on and damaging their fields (Leeson 2013). This was also practised in the new world: in 1713, Franciscan monks in the monastery of St. Anthony of Marañon in the province of Piedade, Brazil, denounced the ants that had invaded their monasteries and sentenced them to leave the territory or suffer excommunication (Lea 1883).

The holy scriptures continue to be the top authoritative book in all aspects of life in the medieval Christendom. The Bible makes ninety-eight references to bees, beetles, fleas, flies, mosquitoes, scorpions, wasps, grasshoppers, lice, locusts, aphids and scale insects, moths and palm worms, as well as hybrids such as the scorpion-tailed locusts (Revelations 12:1–12) and some others whose identity is unclear (e.g., "gnats," Matthew 23:24). Many of these entomological references are concentrated in the Old Testament, especially the Book of Job. Most however do not revere insects, as they are treated with contempt and disregarded as "irrational." Worms (fly larvae) are seen as antithetical exponent of salvation and eternal life due to their association with putrefaction and death. Moths are mentioned many times, all in a negative context as pests of stored goods or cloths (e.g., Job 13:28; Psalms 39:11; Isaiah 51:8; Hosea 5:12; Matthew 6:19; James 5:2). Of the ten plagues of Egypt described in Exodus (7–10), three (the third, fourth, and eighth) directly involve insects. The Book of Isaiah describes hell as a place full of, among other things, spiders and worms (Isaiah 66:24); and the Apocalypse of St. John gives locusts a most demonic and destructive significance, a reputation that left a lasting negative impact on the medieval mindset: insects were relegated to the dark underworld of the demonic, sin and evil. Echoes from this mentality may indeed be the root of some of the modern entomophobias.

But the negative attitude towards insects seems to have gradually changed over time, especially upon the introduction of silk to Europe in the early twelfth century. In fact, similar to the honey bee, the "silkworm" was recognized as a useful insect and illustrated in detail in manuscripts dealing with silk production (Morge 1973). The influence that silk has had on the course of history is unquestionable. Although produced by several insects and spiders, the threads used in making silk fabrics are made by caterpillars of the silk moth (*Bombyx mori*). Silk was first produced in China in antiquity, a monopoly that was maintained until the third century CE. Chinese silk products were imported to Europe for the first time in 120 BCE, and later from Persia in the fifth century, but it was not until the reign of Justinian Augustus in the sixth century CE that Byzantium acquired the know-how of silk production, transforming the Roman Empire from a consumer into a producer of silk (Halsall 1998). Like honey, silk was also a major commodity and demand for it in Byzantium was second only to gold. It was used in ecclesiastical garments, in wrapping relics and crucifixes, and to cover books. Silk shrouds were used in burials, in diplomatic affairs as gifts, and in warfare as displays for allegiance of troops (Muthesius 1992). As such, the production and trade of silk was highly regulated by Rome. Silk producers and merchants were organized into regulated "guilds," and sumptuary laws governed who could wear purple-dyed silk, a symbol of royalty and religious elite. The production of silk and this purple dye, which was obtained from a marine snail, continued to be heavily controlled by the emperor until the tenth century. The Byzantine monopoly on silk production was finally broken upon the fall of Persia to Muslim invaders, who came to the knowledge of silk production and in turn transferred the technology to Sicily and Spain. Silk production later got a foothold in Italy, and by the thirteenth century silk from Europe was competing with Byzantium silk, changing the economy of several regions in Europe (Snell 2017).

Beside the honey bee and the silkworm, few insects managed to gain good reputations in the Middle Ages. In the Old Testament, a spider protected King David when he was hiding in the cave (1 Samuel 24). In Christian books of devotion, ants are often mentioned for their industrious nature, and butterflies and moths are commonly illustrated next to flowers in illuminated manuscripts as marginal decorations, representation of human soul, or next to Virgin Mary representing innocence and virginity (Nazari 2014). Surprisingly, locusts and even bedbugs received some positive credit:

> Now on the first day we arrived at a deserted inn, and when we were at a loss for a bed for John, we saw a droll matter. There was one bedstead lying somewhere there without coverings, whereon we spread the cloaks which we were wearing, and we prayed him to lie down upon it and rest, while the rest of us all slept upon the floor. But he when he lay down was troubled by the bugs, and as they continued to become yet more troublesome to him, when it was now about the middle of the night, in the hearing of us all he said to them: I say unto you, O bugs, behave yourselves, one and all, and leave your abode for this night and remain quiet in one place, and keep your distance from the servants of God.
>
> (Acts of John: 60)

Together with ants, lizards, and hyraxes, locusts were considered one of the four "exceedingly wise" animals (Proverbs 30:25–8). It has been postulated that *Manna*, the biblical food that saved the lives of the Chosen People during the forty years of wondering in the desert, likely refers to the crystalized honeydew made by a certain scale insect (*Trabutina mannipara*) that lives on Tamarix trees (Bodenheimer 1947). Many Catholic or Orthodox saints have insect patronages associated with them; for example, St. Gratus of Aosta (patron of fear of insects), St. Magnus of Füssen (patron of protection from caterpillars), and St. Mawes (patron of protection from all insects). In the year 500, St. Gobneta used bees to deter a band of outlaws terrorizing villagers in York, Ireland. The grave of St. Narcissus of Gerona (patron of protection from biting insects), is said to have been protected from invading French troops in 1258 by a cloud of stinging flies.

With the exception of bees and locusts, insects are hardly ever linked or listed as food items in medieval Europe. At the times of famine, starving masses are reported to be consuming, out of desperation, anything from corpses, bread baked from grape-seed or hazelnut blossoms, fen-roots pressed and dried and powdered, and soup made from dead rats or insects (Montanari 1994).

However, such reports should be taken with a grain of salt since they are likely to be exaggerated or based on hyperbole. The tenth-century Penitential of Burchard, bishop of Worms, established that eating lice, voluntary or not, shall result in a penitence of seven days of water and bread and nothing else (the act of grooming to pick out lice among family members was a common occurrence). In northern Sweden where jewel beetles were considered bringers of good luck and were reputed as protectors of cattle, each year the first beetle seen would be captured, mixed with flour and salt, and given to the leading cow in the farm so that it would be protected from wolves and accidents in the forest.

The Bible condones eating locusts, katydid, cricket, or grasshoppers (Leviticus 11:22), although in religious literature it is not always possible to distinguish between these orthopterans. St. John the Baptist's consumption of locusts and wild honey in the Bible (Mark 1:6; Matthew 3:4) influenced medieval Christian thinkers, but instead of being viewed as an extreme and shocking behavior, resulted in debates about whether locusts should be considered meat or not, since eating any kind of meat for religious men in the Middle Ages was frowned upon (Montanari 1999). Eating locusts has always been common among the people of the middle east (Defoliart 1989). Pliny mentions "Some tribes of the Ethiopians subsist on nothing but locusts" (Bostock and Riley 1855). The fourteenth-century Moroccan traveler Ibn Battuta (1325–1354) talks about locusts as food in eastern Arabia (modern-day Bahrain):

> We came next to Buda, one of the principal villages of Tawat ... There are no crops there, nor butter, nor olive oil; all these things have to be imported from the Maghrib. The food of its inhabitants consists of dates and locusts, for there are quantities of locusts in their country; they store them just like dates and use them as food. They go out to catch the locusts before sunrise, for at that hour they cannot fly on account of the cold.
>
> (Gibb 1929)

Although plagues of locust invasions were frequent and commonplace throughout the Middle Ages in Europe, the reasons behind the formation and movements of swarms of locusts were a mystery shrouded in superstition. They were mainly thought of as divine punishment. In *Naturalis Historia*, Pliny writes about the menacing clouds of locusts reaching Rome from Africa, covering farms and devouring everything. In *De Civitate Dei*, St. Augustine talks about how locusts threw themselves in the sea after destroying many orchards and vegetation in the Roman provinces, and how hundreds of thousands of people (probably exaggerated) died in Messinissa as a result of an epidemic that followed. St. Augustine is also well known for his miracle in the Spanish town of Toledo in 1268, where his prayers caused the invading locusts to rush to drawn themselves in the river Tagus (Alvarez 1959). Today we know that under certain circumstances (unfavorable weather, food scarcity, and high population density), the neurotransmitter serotonin is released, triggering a complex series of behavioral, physiological, and morphological changes that turn solitary grasshoppers into gregarious swarms of locusts that travel many kilometers in search of food.

Other than locust outbreaks, not much else seems to have been known about insect migration in the medieval times, although there is a record of a butterfly migration from 1508 in Calais, France:

> The 23, of Henry the Seventh, the 9. of July, beinge relyke sonday, there was sene at Calleys an innumerable swarme of whit buttarflyes cominge out of the north-este and flyenge south-estewarde, so thicke as flakes of snowe, that men beinge a shutynge in Saint Petar's filde without the towne of Calleys cowld not se the towne at fowre of the cloke in the aftarnone, they flewe so highe and so thicke.
>
> (Turpyn 1541)

Insect metamorphosis, a phenomenon that has captured the human imagination from the dawn of time, was another mystery that influenced medieval religious thinking. In Roman and Greek antiquity, the butterfly (Psyche) was a symbol of soul and transcendent immortal life after death (Blatchford 1889). In the ancient story of Psyche and Cupid (or Eros in Greek), best narrated by Roman writer Apuleius (second century CE), Psyche is an earthly woman whose beauty threatens Venus, the goddess of Love. Venus sends Cupid to take revenge, but Cupid falls in love with her. Venus banishes Psyche to the underworld but she comes back to life victoriously, and is granted immortality by Jupiter so that she can marry Cupid as an equal. Hence, she symbolizes not only the image of the immortal soul, but the anguish and triumph of soul. The Greeks and Romans saw butterflies as personification of Psyche's death and resurrection cycle. Many of the scenes in medieval manuscripts involve *putti*, the little winged children, shooting or otherwise catching butterflies. These putti may be justifiably interpreted as a representation of Cupid chasing his love, Psyche. The abundance of these scenes from an essentially pagan story in Christian religious manuscripts from medieval Europe is rather interesting and demonstrates the continued symbolic representation of butterflies well into the Middle Ages and beyond.

From the tenth century onwards, various insects and arachnids begin to appear on medieval heraldic coats of arms of the European nobility, alongside birds, animals, and monsters. Notable examples include the coat of arms of *Barin La Galissonnière* of France (three butterflies), the *Gresham* family of England (a grasshopper), and the *Barberini* family of Italy (three honey bees).

Three beehives on a coat of arms is a shared motif on crests of several noble families from France and England (Fox-Davies 1909). These family crests were used in standards, armors and shields, emblems and seals, and in jousting and tournaments, and demonstrated social status and aspirations.

Painters in the medieval era incorporated many insect-like chimaeras in their paintings, often with a symbolic or surreal twist. Heavenly beings with insect-like wings are common in religious artwork throughout the Middle Ages and beyond. The scenes of the trial of the tympanum on the west door of the Cathedral of Autun (St. Lazare) in France (1130–1135) are excellent examples of medieval depiction of human figures with butterfly wings. A sculpture in the French Basilique of Vézelay (thirteenth century) depicts a demon with human head and a grasshopper's body, confronting a basilisk. In *L'annunciazione* (1450), Italian painter Fra Angelico depicted an angel with colorful wings like those of a butterfly. In *Temptation of Saint Anthony* (1501), Hieronymous Bosch depicted a grotesque human figure of a cricket. Insectoid beings also abound in his *Garden of earthly delight* triptych (1500): In *Hell*, a demon with the face of a crow and wings of a *Maniola jurtina* (Lepidoptera: Nymphalidae) shoots arrows into the bottom of sinners. In *Fall of Rebel Angels* (1562), Flemish painter Peter Bruegel the Elder also gave an angel the wings of a swallowtail butterfly.

Beside illuminated manuscripts, realistic representations of insects are not uncommon in paintings from the era. A scarce swallowtail (*Iphiclides podalirius*), a pale clouded yellow (*Colias hyale*), and two red admiral (*Vanessa atalanta*) butterflies flutter in the bushes at the background of Pisanello's *Portrait of a princess of the house of Este* (1420). Accurately drawn dragonflies and flies decorate the frescoes in *Les Amants trépassés* in Strasbourg Cathedral, France (1439). In *Jupiter painting butterflies* (1522–1524), Italian painter Dosso Dossi portrays Jupiter, painting butterflies that come to life and fly off of the canvas. Many early Netherlandish and Flemish still-life painters from the fifteenth to seventeenth century commonly included accurately drawn butterflies in their paintings with flowers, fruits, and other objects.

The shift from symbolism to accuracy near the end of the medieval era (fourteenth–fifteenth century), coupled with the renewed interest in the natural world for its own sake and development of new media (e.g., oil paints) to express fine detail, resulted in scientifically more accurate images of animals, including insects, in religious art. By the beginning of the fourteenth century, first-hand observational details of morphology, behavior, and ecology became far more important than before. The *Cocharelli Codex* from the early fourteenth century is a good example of such departure from earlier artistic depictions of insects. Accurately drawn naturalistic images, including those of many insects and caterpillars, decorate the body and margins of this manuscript (Concina 2016). Many practical manuals were written in the fourteenth century on apiculture and agriculture, often critically evaluating, reinterpreting, and correcting classical misinformation, such as *Libre ruralium Commodorum* (Book of Rural Benefits) (1304–1309) by Petrus de Crescenzi where several insects causing food losses, damage, or contamination—including weevils as pests of stored grains and flour—are discussed.

In religious books, such as Books of Hours, butterflies began to take more of a decorative role rather than being symbols of the soul and reincarnation. In paintings, flies were drawn not for their symbolic meaning, but to show the skill of the artist: a highly realistic Sarcophagid fly in Carlo Crivelli's *Madonna and Child* (1480) suggests that the woman in the painting is not Mary the mother of Christ, but Maria Magdalena (Chastel and Mangarelli 1986). By the end of the fifteenth century this trend culminated in Konrad van Megenberg's book *Das Buch der Natur*, the first natural history book with printed illustrations of natural history (Neri 2011). In a radical departure from earlier works, recognizable insects are depicted accurately and in their natural habitat, engaged in their characteristic behaviors. By the turn of the sixteenth century and the beginning of the Renaissance, importance given to empirical evidence, experimentation, and inductive reasoning underlined a new scientific revolution: the natural world began to be used merely as inspiration rather than symbolic or ornamental value. Albrecht Dürer's famous image of the stag beetle (1505) created a movement termed "specimen logic" (Neri 2011), where painters began to stubbornly focus on the materiality of their subject matter. It was these changing attitudes towards nature that inspired later philosophers such as Francis Bacon to employ insect metaphors in explaining scientific theory, and formed the basis on which the first scientific naturalists—such as Joris Hoefnagel, Thomas Muffet, and Maria Sybilla Merian—started their work in the later centuries. Philosophical debates that ensued after the Middle Ages (in the sixteenth and seventeenth centuries) over whether animals, including insects, possess intelligence or could feel pain made way for animal rights movements in Europe and resulted in the creation of organizations such as The Royal Society for Prevention of Cruelty to Animals in 1824.

CHAPTER ONE

Insect Knowledge

MAY R. BERENBAUM

The era of European history beginning roughly in the fifth century and extending through the fifteenth century is generally identified as the Middle Ages or the Medieval Period. In the past, a certain portion of this period has been designated disparagingly as the "Dark Ages," in part to reflect a relative paucity of surviving documents to provide information to subsequent historians, in part to express the perception that the pursuit of academic knowledge fell into abeyance due to widespread political chaos and economic instability, and in part to blame the Christian Church for holding back science. Most medieval historians today, however, do not use the phrase, "Dark Ages," finding the characterization polemical and historically inaccurate. Historians of medieval science share this same view (e.g., Lindberg 1992; Shank 2015). It is certainly the case that the thousand-year timespan of the Middle Ages was far from homogeneous with respect to the status of scholarship. With respect to knowledge of insects, significant advances were made, even if it was not until the mid-eighteenth century that entomology emerged as a discipline with a name.

EARLY MIDDLE AGES (SIXTH–NINTH CENTURY CE): MORALITY AND MYSTICISM AND DIRECT AGRICULTURAL PRODUCTION

Historians differ in how the Middle Ages, or Medieval Period, is defined and subdivided, but according to many there were at least two distinct periods: the Early Middle Ages, spanning the interval from the sixth century to the end of the tenth century, and the Late Middle Ages, from the beginning of the eleventh century until the beginning of the fifteenth century. The Early Middle Ages began in the wake of the collapse of the western Roman Empire, which led to chaos characterized by political unrest, waves of invasion by so-called barbarian tribes, and massive dislocation of populations across the European continent. Thus, not only were traditional trade routes disrupted, causing economic depression, the rise of vernacular languages along with a decline in overall literacy rates interrupted the transmission of knowledge recorded in Latin and Greek during the classical period. Complicating the pursuit of knowledge during this period were waves of epidemic diseases and massive mortality, including the Plague of Justinian, which began in the mid-sixth century and reappeared intermittently throughout the next hundred years, accounting for losses of possibly as much as one-quarter of Europe's population. Agricultural plagues were also problematical, with intermittent outbreaks of locusts with concomitant devastation of crops. Camuffo (1991) documented seven years of locust plagues in Europe in the late sixth century, five years in the ninth century, two in the eleventh century, three in the thirteenth century, fifteen in the fourteenth century and four in the fifteenth century. In the sixth, thirteenth, and fourteenth centuries, locust plagues were followed widespread famines and disease epidemics.

Preservation of scientific knowledge from classical times fell to clerics and theologians, becoming largely the province of monks with scholarly inclinations. These scientific writings often focused not on biology but on astronomy, in part due to the importance of demarcating seasons and years for the purpose of regularizing religious observances. A few books were produced in Latin with biological content, most notably the twenty-volume Latin encyclopedia *Etymologiae*, published by St. Isidore of Seville in 630 CE (Engel and Kristensen 2013). This opus was an effort to chronicle essentially all human knowledge, which encompassed, along with grammar, rhetoric, mathematics, medicine, law, religion (and heresy), languages, portents, the physical world, public works, geology, agriculture, and food, those covered in Book XII: *de animalibus* (beasts and birds).

Much of the information in Book XII derived from texts of the classical period, particularly the first-century *Naturalis Historia* by Pliny the Elder as well as the fourth-century *Hexameron* of Ambrose (so named for its focus on the six days of biblical creation, with insects and other beasts appearing on the fifth and sixth days). Pliny's focus on biodiversity, as was the case for Roman science in general, foregrounded utilitarianism; in his view, organisms merit attention inasmuch as they could provide goods and services to benefit humans. Isidore's *Etymologiae* is unique in the sense that it reflects neither the practical bent of his Roman predecessors nor the spiritual and religious focus of his successors. Insects were discussed primarily in the fifth chapter *de verminibus* (vermin) and the eighth chapter *de minutis volatibus* (tiny flying creatures) (Engel and Kristensen 2013). In total, fewer than three dozen insects were described (only about half the number of birds discussed in the seventh chapter *de avibus*). Morphology was rarely a significant consideration in classifying insects, although Isidore recognized that vermin were distinctive by virtue of the fact that they "do not possess the strong support of a spine, as in serpents" (Barney et al. 2006).

"Vermin," i.e., "animals that are generated for the most part from flesh or wood or some earthy substance, without any sexual congress," were further differentiated by Isidore by what could loosely be called habitat; vermin of the earth, the water, the air, the flesh, leaves, wood, and clothing. Although most "vermis" are species that are today classified as insects, spiders are classified by Isidore as "vermin of the air," the leech "is a water vermin," and scorpions are "land vermin; they are counted more appropriately among the vermin than among the snakes ... it stings with its tail and infuses venom through the bow-shaped wound." Curiously, though, centipedes ("from the large number of feet above") are included among the reptiles. Other vermin worthy of mention were Spanishfly (*cantharis* "that with its burning makes blisters full of liquid as soon as it is applied to the body," clearly *Lytta vesicatoria* or a related meloid); *multipes* (which "rolls up into a ball when it is touched," likely a millipede); slugs (*limax*), silkworms (*bombyx*, from whose tissue "silk garments is accomplished"), and cankerworms (*eruca*), all considered leaf vermin; *teredo*, or wood vermin that eat by grinding; and moth-worms (*tinea*), considered clothes vermin. Among flesh vermin were *hemicranius* (vermin of the head), the maw-worm (*lumbricus,* vermin of the intestines), *ascaris*, *costus*, louse (*peduculus*, vermin of the skin), flea (*pulex*), nit (*lens*), *tarmus* (vermin found in fat, probably *Dermestes*), tick (*ricinus,* a vermin found on dogs), *usia* (a vermin of swine), and the bedbug.

By contrast, "tiny flying creatures" included wasps (*scabrones*), glow-worms (*cicindula*, identified as a kind of beetle), moths (or possibly cockroaches) (*blatta*, "an animal not bear to see the light"), butterflies (*papilio*), locusts (*locusta*), *ciculorum* (grasshoppers "born from the spittle"), cicadas (*cicas*, "born from the saliva of the cuckoo"), flies (*musca*), "dog-fly" (*cynoma*), gnat (*culex,* "the sting of what was said that the blood disappears" [*sic*]), *sciniphes* ("very tiny flies, very troublesome, with their stinging"), horsefly or gad fly (*oestrus* in Greek, *asilus* in Latin, commonly called *tabanus*), *bibiones* ("creatures that are generated in wine," probably drosophilids), and the weevil (*gurgulio*) (Barney et al. 2006). Also found in this chapter are honey bees (*apes*), about which Isidore remarks:

> These are skillful in the business of producing honey, they dwell in homes allotted to them, they arrange their dwellings with a skill that makes no

> mistake, they store the hive from various flowers, and forming their wax-cells, they fill the camp with unnumbered young and they have an army and kings, they make wars, flee from smoke, and are enraged by noise.

With the exception of some gender confusion about queens and workers, these comments likely derive from the direct observations made by Isidore's classical predecessors. Less clear is the disquisition on the reproduction of bees:

> a good many have proved by experiment that these spring from the carcasses of cattle. For in order to create them the flesh of slain calves is beaten, in order that worms may be created from the rotten gore, and these afterward turn to bees. In a correct sense bees (apes) are so called they spring from *boves* as hornets from horses, drones from mules, wasps from asses.
>
> (Brehaut 1912)

This notion of spontaneous generation of insects from decomposing corpses is a holdover from Greek antiquity and actually persisted in some academic circles through the nineteenth century (McCartney 1920), despite Francesco Redi's carefully controlled experiments with maggots and flies in the late seventeenth century rejecting the concept. Confusion may have arisen from the fact that some syrphid flies that are high-fidelity bee mimics (e.g., *Eristalis tenax*) do breed in decaying organic matter.

More representative of the era was *Physiologus*, an anonymous text purporting to summarize ancient Greek and Latin learning, which may date back to the second century CE but which rose to prominence in the fifth century after being translated from Greek into Latin (and subsequently into many other languages). Reflecting the heavy influence of Christianity on learning in this era, the text of *Physiologus* presents aspects of animal biology (or imagined aspects of animal biology) as allegorical lessons for humans; *Physiologus* thus can be considered among the first bestiaries, compendia reporting on the lives of animals, real or mythical, the details of whose natural history are worth noting solely for providing guidance for moral living.

Through its thousand-year history and through translations into at least a dozen languages the *Physiologus* changed in terms of its entomological content but not in terms of its moralistic tone. As Morge (1973) points out, not all versions contained the same information; in his review of medieval entomology, in the version he consulted fifty-six of the total sixty-three chapters are devoted to animals. The 1588 version of *Physiologus*, in Christopher Plantin's *Sancti Patris Nostri Epiphanii, Episcopi Constantiae Cypri, ad Physiologum. Eiusdem in die festo Palmarum sermo,* has twenty-five chapters, twenty of which describe animals of one sort or another. Among insects, ants (members of the family Formicidae) all exhibit social behaviors which can easily be interpreted symbolically as representative of human society; accordingly, they merit special attention in *Physiologus*, one of only three animals discussed in more than a single chapter (Badke 2004). Among the properties of the ant that provide moral lessons:

> The first property of the ant is that they walk in order, each carrying a grain; the ants who have no grain to carry do not try to take the grain others are carrying, but find some of their own. The interpretation refers to the Biblical parable of the wise and foolish virgins (Matthew 25:8) and recommends prudence.
>
> The second property of the ant is that when it stores grain in its dwelling, it divides the grain into two parts so that it will not rot in the winter or germinate if flooded, and the ants will not die of hunger .... the ant is an intelligent animal that can predict the coming of winter, and so works when the sun is hot to gather its harvest while the weather is calm. The interpretation of dividing the grain is that Christians are to divide the Old Testament teaching into two parts: the letter of the law and the spirit of the law. Quoting the Apostle Paul ('The letter kills, but the spirit gives life' [II Corinthians 3:6]) the interpretation says that Christians must obey the spirit of the law, and not be like the Jews who followed the letter of the law and died of (spiritual) hunger. [Figure 1.1]

Many of the details of ant behavior recounted in *Physiologus* appear in earlier classical sources, including Herodotus (fifth century BCE) and Pliny the Elder (first century CE), without the spiritual overlay. These Greek and Roman authorities also describe ants the size of dogs or wolves in Africa that dig up and transport gold, and some versions of *Physiologus*, as well as later bestiaries, reference these fantastic creatures. Another fantastic creature mentioned in discussions of ants in early versions of *Physiologus* is the "ant-lion" (*formicaleon*). This "ant-lion" is a chimerical combination with a front end of a lion and a back end of an ant. The dual nature of this mythical animal was said to symbolize the double nature of every human soul; the moral lesson is an exhortation for every man to walk a single path in life and in prayer.

> Eliphaz the king of the Temanites said, "The ant-lion perished because it had no food." The Physiologus said: "It had the face (or fore-part) of a lion and the hinder parts of an ant. Its father eats flesh, but its mother grains." If they engender the ant-lion, they engender a thing of two natures, such that it cannot eat flesh because of the nature of its mother, nor grains because of the nature of its father. It perishes, therefore, because it has no nutriment. So is every double-minded man; unstable in all his ways ...
>
> (Kevan 1992)

FIGURE 1.1 The ants in the van der Borcht copperplate engraving. Source: Physiologus. http://spcoll.library.uvic.ca/Digit/physiologum/animal/ant2.htm.

Almost a century ago, Weiss (1925) examined the entomological content of *Physiologus* and quoted in their entirety three stories about insects (of the sixty-seven stories in total) from a 1924 translation of the *Physiologus* by one "Mr. James Carlill" published by George Routledge and Sons, Ltd., of London, who in his introduction reviewed the various and sundry versions of *Physiologus* published since 1588. The eusocial life of the honey bee is reported in the context of inspiring men to work collectively for the common good and to obey duly appointed leaders (the latter of which is ironic given the distributed nature of decision making in actual honey bee societies (e.g., Seeley 2010). From *Physiologus*:

> ... bees dwell together in a great company; they go in by one door; they work at the same work; and provide with diligence flying together and

> hastening to do the common toil. And—what is greater than this—they live under a king or prince of like nature with themselves; and they make no beginning of their common work nor hurry to the meadows and flowers until the king who is over them has made the beginning. They choose not the king to this honour but his overlordship comes to him by the rule and custom which the Creator set before them to guide them. Every prince has also a sting but he is not accustomed to wound. Yet, when he turns from his way and stings, he is at once punished for his boldness; for, as soon as he wounds a man with his sting, it remains left in the wound and he himself surely dies from the pain. The bees gather honey out of plants, making use of their mouths; and they entice wax out of flowers, helping also with their feet. And the bees carry away the honey they gather, and out of the wax they build their cells with great skill; and they separate the wax as it were into holes most cunning; and they pour honey into the holes.

Although many attributes of honey bees are correctly detailed—including that the act of stinging is fatal for the worker—other attributes, such as the gender of the queen and workers, had to wait several centuries for the development of tools for dissection and magnification, and ultimately correction. Correct gender assignment, however, had it been known at the time, may well have undermined the religious symbolism embodied by the honey bee. The anonymous authors of *Physiologus* "consider what this means in a spiritual sense"—that the bees who live and work together represent the "holy souls of the sons of the Catholic and Apostolic Church," governed in accordance to divine will "the Archpresbyter," who guards "holiness with his liberty." As for honey, it represents the "spiritual sweetness" derived from the holy flowers of the godly book.

Wasp morphology rather than behavior is the source of the moral lesson; note was taken that the wasp is

> separated or narrowed in the middle of the body, and has not the power to breathe or to draw in the breath with lungs in the same manner as we and the beasts do. But their whole food and nourishment comes from the air; and if thou touchest one with oil, it dies readily, because the greasiness of the oil stops the natural air holes ... The wasps, who are divided in the middle of their bodies and are destroyed as a fire of straw and who perish as soon as oil drops on them, are the rebellious Devils and Fiends, who are divided in the middle of the body that is, by their mutual strife; and they are badly armed for conflict, for, if only thou takest with thee oil, that is the good works full of faith and courage, so mayest thou full well engage in battle with them. For, so soon as they come near the holy oil, their teeth become blunted by

> thy zeal; their air passages become choked by its greasiness; and they are as dead so long as they have not their breath by thy permission. For, if thou persistest not in the anointing or pourest out vinegar while thou slackenest the vigilance of the oil, that is failest in righteousness, then are though like to die, for their life is given to them again as the air-passages open, and they will wound thee with their sting.

The fifty-second chapter presents the dung beetle, whose habit of developing and living in dung provided a convenient and easily graspable metaphor for the heretic: "The dung beetle is declared to be a heretic, sullied by the stench of heresy, and the balls of dung are explained as evil thoughts and heresies" (Morge 1973). Less obvious is what appears to be a reference to fig wasps in Chapter 29 "On the fig tree," in which "gathering figs" is equated with a spiritual experience (Curley 1979); according to Luke 19:4, Zacchaeus gathers figs and scrapes them for holy men but "before they have been scraped there are insects living within the figs in darkness." When the fig is broken open, however, they see the "blazing sun and the moon and the stars" and recognize that they were "in the darkness, sitting in the shadow." The fact that certain figs, including *Ficus carica*, which was established in the Holy Land for thousands of years, are pollinated by gall wasps that remain inside (inedible) caprifigs fruits was known as early as the fifth century BCE (having been described by Herodotus (Goor 1965); this life cycle may well have been the inspiration both for the biblical story and its metaphorical reference in *Physiologus*.

THE CAROLINGIAN RENAISSANCE (NINTH–MID-THIRTEENTH CENTURY) (HIGH MIDDLE AGES)

Although moral lessons could provide enlightenment for the soul, medieval bodies needed attention as well; necessity eventually led to a renewed interest in natural history, for treating diseases and for implementing improvements in agriculture to ward off famines. By the late eighth century, Charlemagne (Charles the Great) reinvigorated education through monastic schools (which would ultimately become Europe's first universities), issuing decrees to restore old schools and found new schools. In view of the fact that monastic or cathedral schools were the principal repositories of knowledge, it is unsurprising that Christian thinking continued to infuse the scientific knowledge reclaimed by translation from antiquity. Compendia from this period, through the ninth century, included *De Rerum Naturis, On the natures of things)* by Rabanus Maurus (776–856), a Benedictine monk who undertook compiling another encyclopedic compendium of all knowledge, this time in twenty-two volumes

(Schipper 1997). Although drawing heavily on the earlier work of Isidore and other encyclopedists, *De Rerum Naturis* (842–847) differed significantly from its predecessors in separating entries on theology from entries on secular subjects, including nature and agriculture (Schipper 1997). Insects and other arthropods are found scattered among the seven chapters in Book 8 on animals. In Chapter 2, *de minutis animantibus* ("on small animals") were *Grillus* (crickets) and *Formica* (ants and ant-lions); in the chapter *de serpentibus* ("on serpents") was *Scorpio* (scorpion); Chapter 3, *de vermibus* ("on worms"), as in earlier works, contains *Areana* (spider), *Cantarida* (likely Spanishfly, *Lytta vesicatoria*), *Multipes* (myriapods), *Limax* (snails), Bombices (caterpillars, including silkworms), *Eruca* (hairy caterpillars), *Teredonas* (wood borers), *Tinea* (clothes moths), *Lumbricus* (earthworms), *Ascaridae* (roundworms), *Tarmus* (probably a dermestid), *Ricinus* (ticks), and *Usia* (a swine parasite). Among the *minuta volatilia* (also called *minutis avibus,* on small birds, in some translations) were *Apes* (bees, with king, worker, and drones), *Scrabrones* (wasps and hornets), *Scarabaei* (stag beetles), *Muscae* (including flies and gad flies), and *Locusta* (including locusts, grasshoppers, and, according to Morge (1973), *Tabanus* and *Drosophila*), along with glow-worm.

Notably, in contrast to the fabulous ant-lions that were literally part ant and part lion, the *formicaleon* of *De Rerun Naturis* refers to what is today considered an ant-lion—i.e., a myrmeleontid. As Rhabdanus writes, "The ant lion is a small animal exceedingly hostile to ants. It hides in the dust and kills the ants that carry provisions. It is rightly called *formicaleon* because it is a lion to ants, if only an ant to all other animals" (Morge 1973).

As the Carolingian renaissance progressed, original observations of nature and practical information became more acceptable forms of scholarship. An example of the practical bent was the publication of the twenty-volume *Geoponika* ("Agricultural Pursuits"), purportedly by Cassianus Bassus (whose authorship has been disputed) for the Byzantine emperor Constantine VII Porphyrogenitus. The tenth-century opus, a mishmash of classical Greek, Roman, and even Zoroastrian sources, was a compilation of information relating to agriculture, including, among other topics, insect-related volumes on "animals and insects injurious to plants" (Book 13) and apiculture (Book 15, "Of natural sympathies and antipathies, and the management of bees"). Many of the pest-control methods recommended are grounded in faith and magic rather than biology; thus, spells and magic are interspersed with cultural practices that are anchored in sound biological principles and that in some cases are still used today. *Geoponika* offers pest-management strategies from classical times, including quite a few from Pliny's *Naturalis Historia* (Books XVII to XIX), including repelling caterpillars by suspending a crayfish or skull of a female horse or ass in the middle of a garden, touching treetops with green lizard gall, or "carrying a toad around the field at night and then burying the creature in the middle of the field in a pot" (Smith and Secoy 1975). Also, from Pliny (XVII, 47) as well, via other classical scholars, was the recommendation "that a menstrual or nubile virgin with bare breast and unbound hair led thrice round a garden hedge caused caterpillars to fall to the ground" (Smith and Secoy 1975).

Among the more realistic recommendations were suggestions for repelling biting insects that may have a biological basis. A mixture of vinegar, oil, and manna was recommended for repelling gnats, particularly when applied to the head and feet. Moore and Debboun (2006) suggest that the acetic acid in vinegar may have suppressed skin bacteria and thereby reduced the production of metabolites used as attractants by some biting pests, including mosquitoes (many species of which have a preference for microbial volatiles emanating from human feet). As well, in *Geoponika* are recommendations for using fumes from burning herbs such as galbanum (*Ferula gummosa*), oregano (*Origanum vulgare*), and bay (*Laurus nobilis*), all of which contain essential oils that today are known to exhibit demonstrable repellency to mosquitoes (Pohlit et al. 2011). Although smoke was the most widely used repellent for centuries, it has the problem of having little residual effect (so it must be produced continuously).

Another practical text relating in part to insects from this era was *Physica*, a pharmacological text written by St. Hildegard (Hildegard of Bingen), a Benedictine abbess who lived from 1098 to 1179 (Throop 1998). Although she published several books detailing her lifelong experience with religious visions, many of which featured animals with symbolic significance, she also published two books on science and medicine inspired by her very practical experience working in the monastery garden and infirmary, although not entirely independent of religious influence; as Throop (1998) points out, "the distinction between the 'medical' and 'visionary' works is not as clear-cut as we might like to think … [T]he so-called medical works were revealed by direct transmission from the Divine, in the same way her more theologically based visions were." Hildegard's text *Subtleties of the Diverse Qualities of Created Things*, called *Physica* in an edition published long after her death, was focused on the medical properties of animals and plants (Czarski 1999). In fact, her detailed writings on plants (in the vernacular German) formed the basis for future European botanical studies (Sabatini 1994). *Physica* consisted of nine books, of which four focused on animals—*Pisces, Volatilia, Animalia*, and *Reptiles*. Insects and other arthropods were scattered across several books. In *Volatilia*, among the insects with medicinal value were the honey bee, the bumble bee, the wasp, the fly, the mosquito, the cicada, and the glow-worm; included among the terrestrial insects (*Animalia*) were ants and fleas. Ants were considered "hot" by standards established by second-century physician-philosopher Galen of Pergamon (according to whom all living things represented a combination of four basic elements: earth, air, fire and water, respectively cold and dry, warm and wet, warm and dry, and cold and wet). Entire anthills were prescribed for phlegm. One remedy for phlegm in the head, stomach, or chest required taking an anthill, complete with ants, and cooking it in water, which should then be poured "over a hot stone" so that steam could be inhaled through the affected nose or mouth ten to fifteen times. For those with a "superfluity of bad humours," preparing a bath with an anthill and fully submerging in the water, exposing only the head, was recommended. Spiders and scorpions (*scorpio*) were placed with the *Reptiles*, sharing with other reptiles "a totally deadly poison that is the death of a human." By contrast, fleas (*Pulex*) and crab lice possessed no medicinal properties but rather were regarded as afflictions; recommended for eliminating fleas from beds was heating dirt on an earthenware tile until it is entirely free of moisture and sprinkling it in the bed (a practice vaguely reminiscent of the use of diatomaceous earth to disrupt and destroy the epicuticular wax layer of pest insects). Honey bees, gnats, and wasps were all considered hot, whereas bumble bees, house flies, tree crickets, grasshoppers, and glow-worms were more cold than hot, galenically. Honey was considered "very hot," to the point that "if a fat person eats it often, it creates mucus and pus in him" although "if a dry and lean person eats cooked honey, he will not be harmed much." Throughout the text are many references to the topical use of honey, as well as its use as a component of herbal beverages and as a coating for a variety of ailments. Notably, honey continues to be used today in many of the same ways.

By the middle of the eleventh century, accompanying the expansion of education in universities, European scholars began to return to texts from the classical era as well as contemporary texts from the Islamic world (e.g., Avicenna and Averroes), translating them into Latin. Historians of medieval science properly place enormous weight on Arab translations of ancient texts, through which the recovery in the tenth–twelfth centuries of (some of) the works of Aristotle and classical scholars was made possible. They also credit the importance of Arabic advances on ancient knowledge. The establishment and proliferation of universities proved critical for science during this transitional period. Although universities may not themselves have provided much new knowledge in the way of information about insects, they did play an important role in encouraging to a degree more naturalistic thinking. The works of Aristotle expanded the hitherto narrowed base of scholarship as they were merged with Christian theology that dominated academic pursuits of the era. Notably, in translating these texts and in teaching students about them, scholars began to supplement them with their own opinions and, importantly, with their own observations.

LATE MIDDLE AGES (THIRTEENTH–SIXTEENTH CENTURY)

By the end of the High Middle Ages, religious symbolism and mysticism began to give way to thinking that allowed faith to be informed by personally acquired knowledge, a critical step in the development of the natural sciences as a freestanding academic discipline. The work of Albertus Magnus exemplified this critical change in scholarship. Albert von Bollstadt (1193–1280) was a Dominican friar (later Bishop) who wrote prolifically across a broad spectrum of subjects, reflecting deep knowledge in all areas of science as well as theology. Among his signal contributions to biology was restoring the zoological writings of Aristotle (translated into Latin from Greek in the mid-thirteenth century by Wilhelm von Morbecke) to curricula and making the case (along with his student Thomas Aquinas) that Aristotle's natural philosophy was not inherently antithetical to Christian philosophy (for much of the High Middle Ages, reading Aristotle was considered heretical). Moreover, in his extensive writings on biology, Aristotle's careful observations in the fourth century BCE were considerably more insightful and reliable than were those of Pliny the Elder, who, in his effort to chronicle all knowledge, rarely took time to verify the information he came across four centuries later.

This is not to say that all of Aristotle's writings were free from errors; in a break from scholarly tradition, Albertus Magnus read Aristotle critically, offered his own interpretations, and on occasion added his own original observations in his commentaries, often correcting Aristotle in the process. As he wrote in *De Mineralibus*, "Natural science does not consist in ratifying what others have said, but in seeking the causes of phenomena" (Bishop 1970).

Among his three dozen or so quarto volumes, Albertus Magnus's *De Animalibus* was the work with the greatest impact on entomological (indeed, all zoological) knowledge. Although nominally a presentation of Aristotle's work on the subject, Albertus Magnus added seven of his own books to Aristotle's nineteen (Morge 1973). The third through seventh of these original books present a compilation of species descriptions of animals known at the time. Of approximately 450 species described, fewer than three dozen are insects, lumped in with other phyla as "small bloodless animals." In a departure from medieval scholarship at the time, external morphology is central to his account of each species. Albertus Magnus noted, among other features, body segmentation of arthropods, the number of legs and wings, differentiation of appendage morphology (e.g., the saltatorial legs of fleas), and internal anatomy (including the length of the digestive tract). Also important in species descriptions, in a major departure from tradition, are details about behavior and ecology, particularly as they relate to humans. Thus, mosquitoes (*Cinifes*) are described as

> flying worms with long legs. They pierce the human skin with a small proboscis. They originate in moisture and are frequent near water. They have a predilection for men and animals that sweat and therefore they are found so much on sleeping persons. In human countries the beds must be covered with special nets to protect the sleepers against their bites.
>
> (qtd. in Morge 1973)

Albertus Magnus also departed from Aristotle in discussing "the intelligence and perfection of animals," including insects and associating specific structures with particular senses (e.g., taste in the proboscis) (Morge 1973).

Many of the details included in species descriptions almost certainly were the result of direct observation, which often placed him at odds with Aristotle. Points of departure with Aristotle were called out as questions (Resnick and Kitchell 2008). Among these questions are those in Book 8 about the honey bee.

- Question 26, "whether the bee can produce honey," about which he writes, "although sweetness is in flowers and fruits, nevertheless these are not a suitable nutriment for bees until they are digested better and converted into honey. This is the reason why the bee naturally makes honey as its nourishment from such redolent things"
- Question 27, as to "whether honey is made in the bee's body or outside it," Albertus Magnus responds that "It must be said that just as a physician grinds up various things so that from them and through them, he will make potions, so too the bee extracts something from flowers and leaves and then grinds and mixes that outside itself and convert it into honey, and later is nourished by the honey made outside itself"
- Question 28, "Whether wax and honey are produced form the same material" is answered with, "it seems not ... honey and wax have different natures, because honey sinks to the bottom of water and wax floats on the top, if they are poured onto water. Therefore, although their efficient cause is the same, it is necessary that they will have different matter ... honey and wax are not produced for the same reason, but honey is produced for nourishment and wax for a house ..."

Other encyclopedic works of the era, as in *De Animalibus*, combined natural history with observations on interactions with humans and offered practical advice. In 1240, Bartholomaeus Anglicus, a Franciscan monk, wrote *De Proprietatibus Rerum (On the Properties of Things)*, a nineteen-volume compendium of all contemporary knowledge, for a lay audience. In addition to classifying insects, he also provided recommendations for pest control; on the flea, he writes,

> The flea is a small worm which plagues people terribly. It feeds on dust. It is very light and flees from danger by hopping, not by running. It grows slowly and is absent in the cold season. It is very fast in summer. The flea is first white but immediately turns black and thirsts for blood. Its sharp bite does not even spare royalty. It is most painful before rain. As a protection against this pest, wormwood, the leaves of the wild fig-tree, and colocynth are recommended.
>
> (qtd. from Morge 1973)

Several of these herbal products (e.g., species of *Artemisia*, the genus to which wormwood belongs) remain in use as flea repellents as part of ethnoveterinary or holistic veterinary practice, albeit without experimental validation (Lans et al. 2008). Thomas Cantipratanus, a Dominican priest and student of Albertus Magnus, spent years compiling his *Liber de Natura Rerum*; although still much in the mode of early encyclopedists, the book nonetheless included practical information, including the use of Spanishfly (cantharides, *Lytta vesicatoria*) against fleas (Morge 1973). Throughout the fourteenth century, practical manuals on apiculture and agriculture began to appear, often based on classical works but in some cases, after the fashion of Albertus Magnus, they critically evaluated and corrected misinformation; such was the case with *Ruralium Commodorum (Book of Rural Benefits)* (*c.* 1304–1309), by Petrus de Crescentii, who was not a cleric but rather a lawyer and judge who retired to write a manual on agricultural practices. Like Albertus Magnus, he relied on classical sources but evaluated, interpreted, and often corrected their writings based on his own observations and experiences. Among the entomological topics covered in *Book of Rural Benefits* were keeping bees and controlling stored product pests.

TRANSITIONAL PERIOD (FOURTEENTH–EARLY SIXTEENTH CENTURY)

As the medieval era was ending in Europe, there was a renewed interest in the natural world for its own sake, facilitated by an awakening of interest in optics, mechanics, and the physical world (Smith 2000). Accuracy trumped symbolic value for many purposes, particularly in illuminated manuscripts and as new media were developed and adopted (including oil paints in the fifteenth century), more detail, more texture, and brighter colors could be achieved than with egg tempera, which by virtue of drying rapidly deprived artists of time required to blend pigments and add fine detail. Crombie (1952) pointed out that without the zoology of the "Western tradition of illustration ... the rise of modern biology would have been impossible."

Insects played an important role in this transition to creating naturalistic (and hence scientifically accurate) images. In religious art, for example, flies continued to figure prominently for their symbolic value as *memento mori* (e.g., Il Vecchietto, *The Four Doctors of the Church Contemplate the Crucifixion*, Carlo Crivelli, *Madonna with Child*), particularly in the late fifteenth-century Dutch Vanitas tradition, e g., symbolizing "the fleeting nature of time and the speed with which death arrives" (Werness 2006). In secular art of the fourteenth and fifteenth centuries and even to some extent in religious art, however, flies were very likely included to showcase the skill of the artist, particularly when used for a trompe l'oeil effect (Chastel and Mangarelli 1986). By the sixteenth century, artists would use the natural world as inspiration and central focus, rather than as symbol or ornament.

That interest was growing in nature for its own sake is demonstrated by images of insects, snails, and other invertebrates in illuminated manuscripts. Throughout much of the Middle Ages, it was a common practice to decorate manuscripts with elaborate initials, borders (marginalia), or miniature illustrations. Although most of the surviving illuminated manuscripts were religious in theme, toward the end of the Middle Ages more secular works were illuminated outside monasteries by lay professionals in so-called scriptoria. Nazari (2014) surveyed hundreds of manuscripts spanning the period from 1280 to 1540 and found approximately 270 that included images of Lepidoptera. Remarkably, among these are about thirty depicting people, monkeys, putti, or other mythical beings, pursuing and capturing butterflies. Butterfly collecting appears at least as far back as the mid-fourteenth century, in *The Romance of Alexander*, a Flemish illuminated manuscript. The accuracy of the images suggests that drawings were made from real specimens.

The Cocharelli Codex, from the early fourteenth century, also represents a departure of sorts from earlier artistic depictions of insects (Concina 2016). The text is divided into two parts—a prose essay on vice and virtue, written for the edification of the sons of a Genoese merchant and based on stories from the (anonymous author's) grandfather, Pelligrino Cocharelli and a Latin history in verse of Emperor Frederick II's exploits in Sicily, also based on recollections of the grandfather. The text in both sections is enlivened by naturalistic images, primarily of insects, along the margins (and playfully, as well, with caterpillars as line fillers) (Figure 1.2). At first glance, the images might be thought to be entirely unrelated to the content of the text, but an analysis reveals that the illustrations accompanying the historical verse actually convey information that reinforces the historical lessons.

FIGURE 1.2 Marginalia Insects. © British Library Board, Add. 28841, f. 4v (11 × 17 cm). http://www.bl.uk/catalogues/illuminatedmanuscripts/ILLUMIN.ASP?Size=mid&IllID=24300.

Our surprise becomes amazement at the animal scenery scattered in the margins surrounding the text of the historical poem. These margins are literally invaded by a diversified fauna, mainly entomological. This zoological gallery unrelated to the text has long been considered strictly ornamental. However, a recent in-depth analysis reveals that animals were not at all

> an innocuous setting, but consisted of natural science lessons in pictures that would awaken the sense of observation of nature and the awakening of curiosity with regard to the concrete. In addition, the distribution of animals according to their habits, figurations obviously educational, the location of immature stages and adult stages, highlight a perfect knowledge of the Treatise of Zoology by Aristotle. For example, we can observe the animals identifiable on two pages borrowed from the historical narrative ... At the top of the first page, the margins of which are lined with ornamental ivy, two profiles of Scarabé beetles (*Oryctes nasicornis*) face each other at the top of the page. Next to it is placed an aquatic Hemiptera with posterior swimming paws *muni de pattes*: a notonect. In the right margin, at the top and at the bottom, are represented two Pentatomid Hemiptera, bugs with typical geometric forms. The artist has represented the top insect with anterior coriacous wings resting on the back while the lower one is pedagogically presented with the fore wings apart to reveal the second pair of membranous wings. This type of didactic demonstration is taken up on the *Oryctes* of the basal margin: the second pair of unobstructed wings shows well its network of ribs. The beetle is framed by two butterflies of the same species but one, drawn in dorsal view, has spread wings while the other, seen in profile, has the wings at rest. In their typical coloring it is easy to recognize male Arctiidae: *Utethesia pulchella*, a migratory southern species. We also note in the top of the entrecolonne, an immature form of stink bug with short winglets revealing the flattened dorsal surface of the insect.
>
> (Bistch 2013)

Thus, by the beginning of the fourteenth century, details of morphology, behavior, and ecology, learned through first-hand observation, became far more important than they had been for much of the preceding millennium. This transition is also evidenced in the work of the otherwise anonymous Master of the Brussels Initials in the late fourteenth century (Hutchinson 1974). The borders he created for manuscripts are replete with naturalistic images with accurate anatomical details. One such work was a Book of Hours (British Museum, Add. M29433), a popular type of literature at the time that contained short services that could be sung or spoken. A diversity of insects are depicted with sufficient detail as to be recognizable to species; the Master of the Brussels Initials was particularly partial to aposematic insects such as the firebug (*Pyrrochoris apterus*) (e.g., fol. 192v M border contains a firebug *Pyrrhocoris apterus,* along with an apollo butterfly *Parnassius apollo* and a tortoiseshell butterfly *Nymphalis polychloros (Aglais urticae?)* (Hutchinson 1974)). Hutchinson (1974) observed that the Master of the Brussels Initials often depicted moths with an extended proboscis and surmised that such images were most likely based on real specimens or on drawings of real specimens available to the artist.

By the late fifteenth century, in 1475, this trend toward realism culminated in a landmark work published by the fourteenth-century scholar Konrad von Megenberg—*Das Buch der Natur*, a translation of *Liber de Natura Rerum* by Thomas Cantipratanus, written a hundred years earlier (Figure 1.3). Although *Das Buch der Natur* was published around 1348, changes in publishing practices in the mid-fifteenth century greatly expanded its influence. Printed books with woodblock printing, along with printed books produced with Gutenberg's moveable type, made works affordable to a much wider audience of readers by this time, and von Megenburg's translation of the earlier work into the vernacular German made the information in *Das Buch der Natur*, the first natural history book to be published in German, considerably more accessible as well. It was also noteworthy by virtue of its twelve folio plates of woodcut illustrations, the first printed illustrations of natural history (Neri 2011). Insects are recognizable by their morphology to order and often to species; more importantly, many are depicted in their natural habitats, engaging in typical behavior, or in characteristic ecological interactions—in one such woodcut, a butterfly is shown nectaring on a flower, a spider is sitting in the center of its web, bees are depicted flying in and out of skeps, and ants are shown walking around on an anthill. Although the images are too crude to distinguish what the ants are carrying—they could be seeds or pupae, given that there are no details beyond an outline—this artistic rendition of the actual lives of the subject is a radical departure from earlier works in which insects serve only as symbols for religious beliefs or as colorful ornaments for religious texts.

The turn of the sixteenth century is considered by some as the definitive end of the Middle Ages and the beginning of the Renaissance and the subsequent Scientific Revolution of the seventeenth century, reliant upon empirical evidence, experimentation, and inductive reasoning. In 1505, Albrecht Dürer famously created an image of a stag beetle, with fine details of its external anatomy meticulously depicted in watercolor and gouache, that, with other such images, created a movement Neri (2011) calls "specimen logic," with a focus on individuals "stubbornly insistent on the materiality of their subject matter." With specimen logic, art and science would merge in using the natural world as inspiration (as Dürer wrote, "art is omnipresent in nature, and the true artist is he who can bring it out"—Foottit and Adler 2016). It is indeed apt that, when Francis Bacon, among the men credited as a founder of the scientific revolution of the seventeenth century, introduced his new method of conducting scientific investigation through experimentation and inductive reasoning in *Novum Organum Scientiarum* (*A New Instrument of Science*) in 1620, he deployed insect metaphors to explain it—to a new purpose perhaps, but still in a way that suggests continuity with the thinking of the Middle Ages.

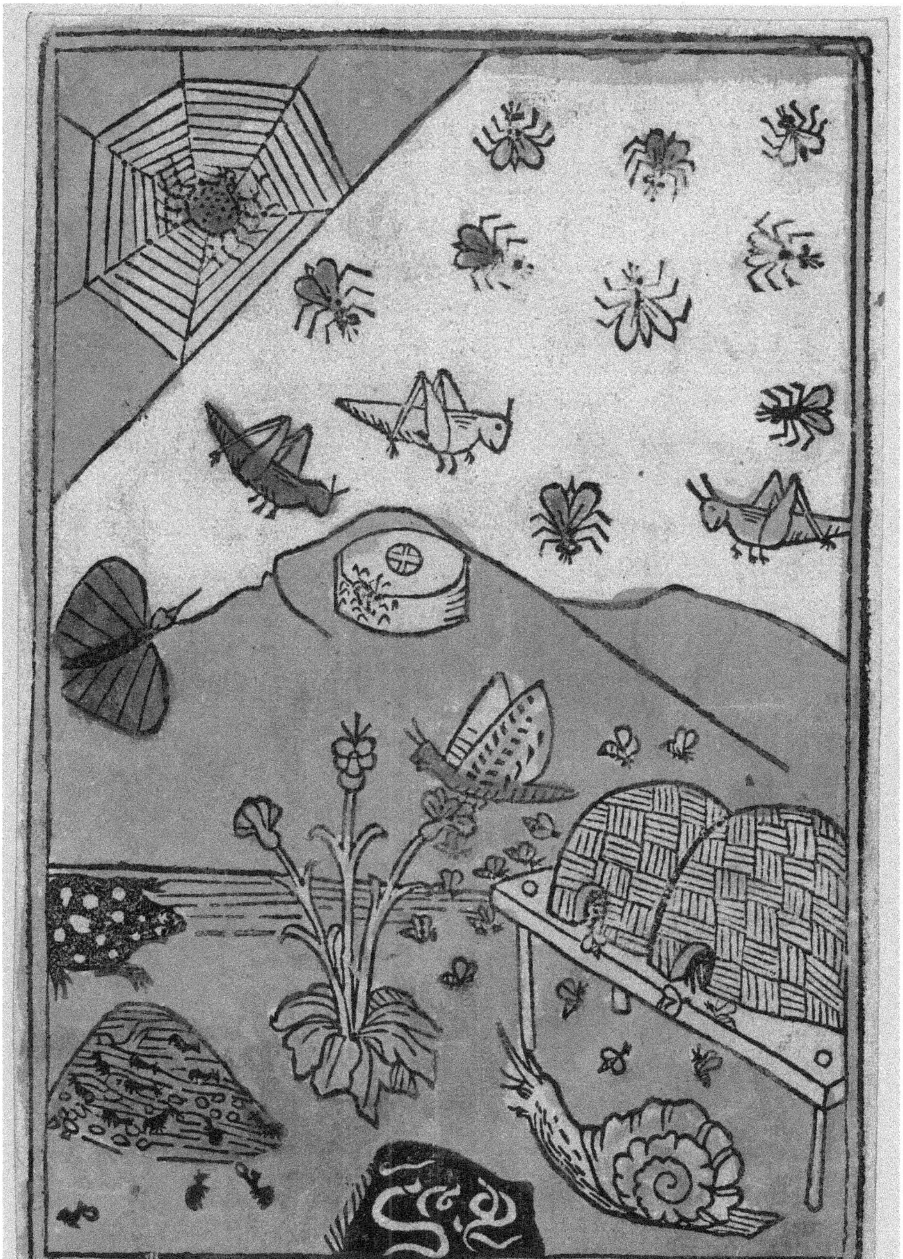

FIGURE 1.3 Woodcut from Konrad von Megenberg, "Buch der Natur," printed by J. Bämler, Augsburg, 1475–1481. British Museum. http://www.britishmuseum.org/research/collection_online/collection_object_details.aspx?objectId=1353907&partId=1.

Bacon explained the different approaches to scientific inquiry metaphorically with ants, spiders, and bees, who respectively collect, reason independently, and glean and process information (Berenbaum 2018). Indeed, the Middle Ages were not lacking in careful observers of insects, and insect knowledge, and it was on this foundation that the discipline now called entomology was built.

CHAPTER TWO

Insects and Disease

EILEEN REILLY

The Medieval Age was a time of great change—culturally, socially, economically, climatically—across the globe. Major population movements, new long-distance trading routes, and invasions put pressure on both land and people. Inevitably, famine, war, and disease followed in their wake and took their toll. The medieval economic historian, Bruce Campbell, coined the phrase "The Great Transition" for the fourteenth century, in particular, because of the combined impact of commercial recession, war, climate change, and the Black Death on the peoples of Eurasia (Campbell 2016). Insects undoubtedly played an important role in the movement of diseases, which took advantage of both trade and war to produce epidemics in new population groups.

This chapter will look at the insects and other arthropods that either spread or maintained diseases during the Medieval Age in both humans and animals. It will present examples from the period of the devastating effects of disease, perceptions of insects during this time from documentary sources and awareness (or lack thereof) of the causes of major illness. Examples from across the globe, where available, are used to illustrate these points.

INSECTS AS VECTORS OF DISEASE

Today, we know that the major insect orders associated with disease spread in both humans and animals are flies (Diptera), especially mosquitoes, blackflies, house and stable flies, horseflies, tsetse and ked flies, and biting midges; fleas (Siphonaptera) and lice (Phthiraptera) (Table 2.1). Bugs (Hemiptera), such as the kissing bug, are associated with Chagas' disease in the Americas, although the better-known bedbug (*Cimex lecturlaris*), is not a known vector of any disease (Delaunay et al. 2011). In addition, non-insect invertebrates, such as ticks and mites, are known vectors of diseases and will be discussed here also in the context of historical outbreaks (Slavin 2016). Our understandings of insects as vectors of disease is primarily a product of post-medieval and early modern scientific discovery (Gubler 2009; Newfield 2017). For the most part, people living in the Medieval Age were unaware of the causal linkage between insect bites and disease (see the Discussion and Conclusions section at the end of this chapter).

TABLE 2.1 Principal orders of insects and other arthropods that transmit diseases in animals and humans (adapted from Hopla 1994; Gubler 2009). Diseases underlined are discussed in the text.

Insects and other arthropods	*Pathogens transmitted*	*Notable diseases*
Insecta		
Anoplura (sucking lice)	Viruses, rickettsiae, skin fungi, nematodes	Epidemic typhus, murine typhus, Relapsing fever, plague?
Mallophaga (chewing lice)	Skin fungi, cestodes, nematodes	
Hemiptera (bugs)		
Cimicidae (bat bugs/bed bugs)	None	
Triatominae (kissing bugs)	Protozoa	Chagas' disease
Diptera (flies)		
Culicidae (mosquitoes)	Viruses, bacteria, protozoa, nematodes	Malaria, Yellow fever, West Nile virus, various encephalides incl. Eastern equine encephalitis
Psychodidae (sandflies)	Viruses, protozoa	
Simuliidae (blackflies)	Protozoa, nematodes	
Ceratopogonidae (biting midges)	Viruses, protozoa, nematodes	Bluetongue
Tabanidae (horse and deer flies)	Viruses, bacteria, protozoa	
Muscidae (house flies etc.)	Bacteria	
Calliphoridae (blow flies)	Secondary infection	
Sarcophagidae (flesh flies)	Secondary infection	
Glossinidae (tsetse flies)	Protozoa	Sleeping sickness
Hippoboscidae (louse flies)	Protozoa, nematodes	

Insects and other arthropods	*Pathogens transmitted*	*Notable diseases*
Siphonaptera (fleas)	Viruses, bacteria, rickettsiae, cestodes	Plague, murine typhus
Arachnida		
Ixodida (ticks)	Viruses, bacteria, rickettsiae, protozoa, nematodes	Encephalides, Lyme disease, various types of typhus
Psoroptidae (mange mites)	None	Sheep scab
Trombiculidae (chigger mites)	Viruses (?), rickettsiae	Scrub typhus or Japanese river disease

Insects spread disease in two ways, i.e., mechanical or biological transmission (Gubler 2009). The first involves direct contact between the insect and the disease pathogen and simple transfer on the feet or mouthparts of the insect to a human or animal. Various bacteria and viruses are transferred via direct fecal-oral transmission and primary culprits include house and flesh flies. The house fly was once known as the "silent destroyer" and was implicated in the spread of many virulent diseases throughout history, notably cholera, tuberculosis, infantile gastroenteritis, and possibly diphtheria and smallpox (Ross 1914). In addition, stable and horse flies can mechanically transmit blood-borne diseases from open sores as they move from host to host. This mechanical transmission of disease by insects, while clearly understood now in the context of modern epidemiology and hygiene practices, is less easy to explicitly demonstrate during the Medieval Age. It can, however, be inferred through an understanding of the living conditions of the time and, to some extent, archaeoentomological evidence (e.g., Smith 2012; Reilly 2018) (see the Discussion and Conclusions section).

The second method of transmission, particularly important in the context of the Medieval Age, is biological transmission (Gubler 2009). This is where the pathogen undergoes some form of biological change within the insect, e.g., simple multiplication of the pathogen or a full developmental cycle or the pathogen is transferred from parent to offspring via eggs. The newly hatched larvae are thus infected with the disease and pass it on in subsequent developmental stages. Many of the important diseases that impacted on the peoples of the Medieval Period across the globe fall into this category of transmission.

In the context of the case studies that will be discussed in this chapter, a brief outline of the important insects and other arthropods, which vector disease, is warranted. Some are better known than others, while some have had a more insidious overall impact on human and animal health during the Medieval Age than previously understood (see Table 2.1).

FLIES

Flies are by far the most important insect group in terms of disease transmission today, primarily due to the mosquito family (Hopla 1994; Gubler 2009). Of the thousands of species of mosquitoes in the world, hundreds are pests of livestock, poultry, and humans and many are important vectors of pathogens. Viruses such as encephalitis, bacteria such as anthrax, and parasites such as malaria are all transmitted by mosquitoes to animals and humans via both biological and mechanical transmission.

The nuisance value of flies to humans and animals was well recognized by writers and artists of the Medieval Age (Figure 2.1). The Arabic author

FIGURE 2.1 Detail of a marginal painting of flies surrounding a dog, from the "Maastricht Hours," Netherlands (Liège), first quarter of the fourteenth century. © British Library Board, Stowe MS 17, f. 48r. http://blogs.bl.uk/digitisedmanuscripts/2014/08/bugs-in-books.html.

of Kitāb Al-Hayawan, Al-Jāhiz, writing in the ninth century CE, gave a vivid account of being plagued with biting flies, possibly mosquitoes, on the banks of the River Basra, declaring that "Flies such as these that live in the marsh, grasses and orchards have a bite second to none" (Jones 2013). The mid-tenth-century farming manual *Geoponika*, created in Byzantium, advised farmers that "stretched horsehair across the door and through the interior of the house destroys mosquitoes and prevents them from entering" and "if you soak a sponge in sharp vinegar and hang it at your head and feet mosquitoes will not bite you" (Dalby 2011).

The presence and prevalence of malaria in the Medieval Age has been somewhat neglected by researchers, despite the fact that malaria or malarial-like illnesses were described by many writers in antiquity from China to India, the Near East to North Africa and Roman Italy (Cox 2010). However, research by Dobson, Sallares, and most recently by Newfield is casting new light on the prevalence of malaria in Europe and North Africa during this period (Dobson 1980, 2003; Sallares and Gomzi 2001; Sallares 2002; Newfield 2017) and will be discussed below. In addition, the possibility that mosquitoes were responsible for another significant disease outbreak in the late eighth century CE, this time in animals, will be highlighted.

FLEAS

Fleas are small flightless insects that live as external parasites on mammals, birds, and humans, consuming the blood of their hosts (Whiting et al. 2008). They primarily constitute a nuisance, causing itching at the point of their bites, but their impact can also be more malign in terms of disease, parasite, and bacterial transmission. In the tenth-century CE *Geoponika*, mentioned above, advice is given on how to rid one's house of fleas (Dalby 2011):

> Dig a hole, grind oleander leaves and place in it, they will all gather there. Otherwise, soak the floor repeatedly with amorge, then grind wild cumin and mix with waters, and grind 10 drams of squirting cucumber seed and add to the water, sprinkle this in the room and you will make the fleas split.

A recognition perhaps even then that some fleas do not live permanently on their host but also in bedding, clothing, floors, and even thatch.

Fleas have been recorded in archaeological sites from all across the world, notably the human flea (*Pulex irritans*) and the cat flea (*Ctenocephalides felis*), with the former particularly prevalent in Medieval Age sites from Europe and the North Atlantic region (Buckland and Sadler 1989; Reilly 2003; Bain 2004; Panagiotakopulu 2004; Forbes et al. 2013; Huchet 2015).

Many fleas are host-specific, i.e., they do not move from one type of animal to another, but others can parasitize a number of different hosts (Whiting et al. 2008). For the Medieval Age, the assumed role played by rodent fleas (notably, *Xenopsylla cheopis*) in the spread of the bubonic plague bacillus *Yersinia pestis* during the fourteenth century CE, and probably the earlier "Justinian Plague" between sixth and eighth centuries CE, is perhaps one manifestation of this. These constitute some of the most devastating disease epidemic events in human history, certainly among the best known and most written about (Benedictow 2004). However, there remains the conundrum that *X. cheopis* has not been recorded from Medieval Age archaeological sites across Eurasia, despite hundreds of excavations in locations with excellent preservation (Reilly 2003, 2018; Smith 2012; Huchet 2015). Could the human flea, therefore, have been the primary vector of plague at this time, as well as many other infectious diseases? Discussion of this issue and the overall impact of fleas in the Medieval Age is outlined below.

LICE

Lice are small wingless insects categorized as either "sucking" or "chewing/biting," which cannot readily live independently from their host (Hopla 1994). The sucking lice are the most important in terms of disease transmission for humans and animals with diseases such as epidemic typhus, trench fever, and Relapsing fever in humans, and swinepox virus in pigs, transmitted by these parasites (Boutellis et al. 2014). Humans are parasitized by two types of lice: *Pediculus* on the head and body (two sub-species), and *Pthirus* in the pubic area. They have co-evolved with humans for many millions of years and archaeological evidence for their presence has been recorded all over the world, from ancient Egypt, to New World pre-Columbian mummies to eleventh–fourteenth-century Norse and Inuit sites in Greenland (Araujo et al. 2000; Bain 2004; Huchet 2015).

Lice and the act of de-lousing have been portrayed in art and described in literature in the Medieval Age, notably the "biblical plague of lice" image in the Golden Haggadah, from the mid-fourteenth century (Figure 2.2).

De-lousing (Figure 2.3), although generally viewed as a dirty and secretive activity, was occasionally written about in vivid terms by medieval writers. One such example comes from the French village of Montaillou written down during a fourteenth-century inquisitional testimony (Smith 2007):

> Pierre Clergue had himself deloused by his mistresses ... the operation might take place in bed, or by the fire, at the window or on a shoemaker's bench ... Raymond Guilhou also deloused the priest's mother, wife of old Pons

Clergue, in full view of everybody in the doorway of the "ostal", retailing the latest gossip as she did so. The Clergues, as leading citizens, had no difficulty in finding women to relieve them of their insect life ...

FIGURE 2.2 Image showing the biblical plague of lice, Golden Haggadah, Catalonia, second quarter of the fourteenth century. © British Library Board Add. MS 27210, f. 12v. http://blogs.bl.uk/digitisedmanuscripts/2012/02/hebrew-illuminated-manuscripts-online.html.

The role that lice played in the transmission of disease during the Medieval Age is difficult to ascertain, however. Historical records for outbreaks of "typhus"-like conditions will be assessed, as well as some interesting theories on the origins of typhus in medieval Europe directly related to the movement and mingling of human parasites across the globe.

BUGS

Bugs are relatively large insects (in comparison to fleas/lice), several of which are blood-sucking ectoparasites (Hopla 1994). The bedbugs *Cimex lecturalis* (global species) and *C. hemipterus* (tropical species) feed on humans but also other animals, including chickens. It is thought that *Cimex* species were parasites of bats first but came into contact with humans in cave environments. They have been found in archaeological sites from Egyptian times onwards, but Kenward has noted the rarity of finding bedbugs in Old World sites dating from the Roman and Medieval periods (Kenward and Allison 1994). One illustration from late fifteenth-century Germany appears to show bedbugs crawling across a bed, accompanied by advice on how to rid oneself of them (Figure 2.3).

FIGURE 2.3 Images of bedbugs and de-lousing in "Hortus Sanitatis," Jacob Maydenbach, Mainz 1485, first natural history encyclopedia. Credit: Wellcome Collection. https://wellcomecollection.org/works/yqpcqup3.

Their role as disease vectors has been discounted but they constituted a nuisance to human health through their bites, which can result in dermatitis and secondary infections (Bain 2004).

One family of bugs, however, that is implicated in disease transmission are the Triatomine bugs or kissing bugs, found throughout the entire Amazon basin and other parts of South America. Their role in vectoring Chagas' disease is well attested and there is some suggestion that this pathogen impacted pre-Conquest civilizations in South America (Coimbra 1988). Contemporary documentary sources are not available so further discussion of this topic in relation to the Medieval Age is not possible.

TICKS AND MITES

Ticks and mites are not insects but members of the Spider family (Arachnids). All ticks are obligate blood-sucking ectoparasites of birds, animals, reptiles, and/or humans at some stage of their life cycle and many are vectors of diseases, such as Lyme disease, various forms of encephalitis and typhus (Hopla 1994). Specific linkages between ticks and disease spread in the Medieval Age is difficult to come by. However, there is some evidence in the historical record for the impact of mites on animal populations in the thirteenth century CE (Slavin 2016), which will be discussed below.

CASE STUDIES FROM THE MEDIEVAL AGE

Mosquitoes: Malaria in Early Medieval Europe

Malaria is caused by Plasmodia parasites and is transmitted to humans from the mosquito genus *Anopheles*, who also "host" the parasites in their body (Newfield 2017). Three types of malaria are of interest historically, i.e., *Plasmodium vivax, P. malariae*, and *P. falciparum*. The first two are considered "mild" forms of malaria, with *P. falciparum* the deadliest form. Mediterranean Europe appears to have had a long history with the *vivax* form of malaria, judging by genetic studies, while *falciparum* made occasional deadly outbursts from Africa from classical antiquity onwards (Hoffmann 2014). To be successful, malaria needs both a reservoir of humans who have the disease and suitable mosquitoes to transfer it from one human to another. Mosquitoes and Plasmodia also have narrow habitat and temperature requirements, which can both limit and expand their range, depending on climatic and other environmental factors (Hoffmann 2014; Newfield 2017). *Anopheles atroparvus*, for example, the mosquito species associated with many historical outbreaks of malaria, favors stagnant or brackish water (coastal marshes, lagoons, canals, ditches), while other species prefer fresh water, river plains, and lake margins.

Newfield (2017), in his detailed and scholarly assessment of malaria and malaria-like diseases in early Medieval Europe, concludes that references to "cyclical fevers" in documentary sources (the way malaria manifests itself in people) and palaeoscientific analyses (aDNA) constitute the best forms of evidence for the presence of the disease at this time. In this way, a definite pattern of disease outbreak can be seen in Merovingian France (450–750 CE) and on into the Carolingian era (750–887 CE). Various references to cyclical fevers in Italy, France, Germany, and England occur from the ninth century CE onwards, including the possibility that Charlemagne himself suffered from malaria in later life (McKitterick 2008). It appears to have been *endemic* (i.e., regularly found in a local area or local population) in the Frisian area from the ninth century when writers referred to "Italian fevers" along the Rhine, associated with settlement of newly drained marginal wetlands (Hoffmann 2014). Newfield (2017) argues that malaria probably became more prevalent from the onset of cerealization in the Carolingian period through deforestation, arable expansion, coupled with increased riverine and coastal sedimentation, which may have helped to expand areas occupied by the *Anopheles* mosquitoes. This is in contrast to widely held views that it was more prevalent in the aftermath of the decline of the Roman Empire, due to the breakdown in water management, a generally wetter climate and wetland expansion. Hoffmann (2014) notes that *P. falciparum* became a scourge in coastal Sicily from the thirteenth century when lagoons were turned into marshes by the deposition of soil eroded from over-farmed arable land. He notes other writings referring to "fevers" northeast to Jutland, where marine incursions and flooding carried more salt-water into marshes, inhibiting mosquito competitors of *A. atroparvus*. In general, early medieval European endemic zones for malaria would have occurred in overcrowded riverine and coastal settlements and well-populated agricultural communities. Indeed, it is possible that through a combination of primary infection, susceptibility to secondary infections and malnutrition, endemic malaria took a greater toll on Medieval Age populations than occasional plagues. Productivity in farming regions suffering from malaria would have been limited, possibly leading to abandonment of some labor-intensive activities such as viticulture (McCormick 2002). The impact, therefore, of the mosquito as a vector and maintainer of disease in the Medieval Age is undoubtedly far greater than hitherto understood.

Mosquitoes Again? The "Equine Epidemic" in Charlemagne's Horses

Anecdotally, the mosquito may also be responsible for another disease outbreak during the Medieval Age, but one that affected animals, not humans. In 791 CE, a contingent of horses in Charlemagne's army that were passing along the marshy banks of the Danube during the Avar Campaign (in what is today lower Austria) contracted a disease and died, apparently in great numbers (Gillmor

2005; Newfield 2015a). The Revised Frankish Annals described the epidemic or *epizootic*, to give it its correct name (i.e., specifically a disease outbreak in animals), as so virulent that "... scarcely a tenth part of some many thousands of horses is said to have survived" and may have had long-term repercussions for Charlemagne's ability to wage war for a number of years afterwards due to the shortage of seasoned warhorses (Gillmor 2005). Certainly, Gillmor argues that it needs to be taken more seriously as the cause for Charlemagne's apparent two and a half years of inactivity between late 791 and mid-794 CE.

Unfortunately, contemporary accounts do not provide enough evidence to give a definitive cause for the disease, with some writers suggesting a disease known as equine strangles, caused by the *Streptococcus equi* bacillus (Hyland 1994). However, others have doubted whether this disease, which requires direct contact between infected and non-infected animals, could have killed as many horses in their prime as that described in 791, as it generally affects much younger or older horses. Eastern Equine Encephalitis (EEE) has been suggested due to its high morbidity rate and the fact that the marshy banks of the Danube were undoubtedly mosquito-infested (Gillmor 2005). The primary hosts of the disease today are *Culiseta melanura* and *C. morsitans*, but these are American species of mosquito, where the disease originates (Arrigo et al. 2010). It is known to be vectored by the most common species of mosquito in Europe, *Aedes vexans*, but EEE's arrival in Europe is considered relatively recent as a result of animal trade in exotic birds and reptiles (Durand et al. 2013). Therefore, could EEE really be the cause of this famous equine epizootic in the early Medieval Age? In the absence of palaeoscientific analysis, i.e., aDNA from horse skeletons of the period it is impossible to be definitive about the cause of such a devastating loss of animals.

Fleas and Bubonic Plague: The "Justinian Plague" and the Black Death

Much has already been written about the most famous insect-vectored disease epidemics in history—the "Justinian Plague" and the Black Death—and a detailed re-telling or analysis is beyond the scope of this chapter (e.g., Gottfried 1983; Benedictow 2004; Crawford 2007; Hoffmann 2014; Campbell 2016). This chapter will, therefore, focus on the evidence for the assumed insect vector, the flea, and its host, the Black rat.

Although there has been much debate over the years as to the exact cause of the Black Death and the Justinian Plague (see Twigg 1985; Scott and Duncan 2001; Cohn 2002; Campbell 2016 for a review of the debate), recent scholarship has concluded that it was most likely caused by bubonic plague (*Yersinia pestis*) and related forms of the disease. The debate centered around the view that the disease as known today manifests itself somewhat differently to the descriptions from the Medieval Age, particularly the rapidity with which it killed and the speed with which it apparently moved (Figure 2.4). There were also questions as to whether the rodent flea (*Xenopsylla cheopis*), the primary host of the disease with its preference for tropical climes, could survive in more northerly climes, and whether the flea's assumed primary host (particularly in the context of the Black Death), the Black rat (*Rattus rattus*) ever became established in Europe (Panagiotakopolu 2004; Huchet 2015).

FIGURE 2.4 The burial victims of the plague in Tournai. Detail of a miniature from "The Chronicles of Gilles Li Muisis" (1272–1352), abbot of the monastery of St. Martin of the Righteous (Bibliothèque Royale de Belgique, MS 12076–77, f. 24v). Credit: Wikimedia Commons.

The question as to whether *Yersinia pestis*, the plague bacillus, was the cause of both disease outbreaks has been greatly aided by recent palaeoscientific analyses, which detected the disease in skeletons from the sixth century CE and from the fourteenth century CE in France, Germany, the Netherlands, Great Britain, and Italy (Raoult et al. 2000; Haensch et al. 2010; Harbeck et al. 2013). Although the Justinian Plague began in the 540s CE, waves of outbreak continued into the eigth century CE. It appears to have spread from Africa or possibly the Arabian Peninsula across the Mediterranean and into Europe (Benedictow 2004). Notably, writers of the time describe it as something "outside their previous experience" (Hoffmann 2014). It eventually reached Ireland and Great Britain by the mid-sixth century and further outbreaks were recorded there in the seventh century. It is speculated that it had a significant impact on population movement and economic development in regions of Europe during this time. There has even been some suggestion that defense against its spread led to the distinctive dispersed settlement pattern characteristic of early medieval Ireland from the sixth century onwards, in contrast to the larger clustered settlements seen elsewhere in Europe at this time (Hoffmann 2014; O'Sullivan et al. 2014). Recent analysis of the *Yersinia* genome shows the strain of *Y. pestis* associated with the Justinian Plague was different from that which caused the Black Death and, indeed, that several strains of the disease were active at the same time during the fourteenth century (Parkhill et al. 2001; Hinnebusch 2005; Cui et al. 2013; Wagner et al. 2014). The changing nature of the disease partly explains its virulence during the Medieval Age as it was perhaps more difficult for people to develop immunity from several different strains at once. Any immunity built up in populations from earlier outbreaks would also have been ineffective against these new strains. Coupled with the fact that people of this time may have had compromised immune systems due to other pathogens, intestinal parasite load and malnutrition, it is not surprising that a "new" disease introduced to new populations could cause such devastation.

But what of its vectors? Bioarchaeological evidence has shown that the Black rat did indeed live in northern Europe from as early as Roman times, although it may have died out in places like Britain and Ireland and been reintroduced regularly throughout history as a result of trade (Benedictow 2004). Skeletal remains of the Black rat from Great Britain, Germany, Poland, and as far north as Sweden, dating to the Medieval Age, also suggest it was not as temperature-dependent as people once thought. Further, the apparent seasonality of the disease in some of these regions—with greatest numbers of deaths occurring from summer to late autumn—may be directly related to the greater activity levels of both rats and their fleas during the warmest months of the year, with activity suppressed during winter and early spring (Benedictow 2004; Campbell 2016).

If the Black rat was able to inhabit northern Europe during the Medieval Age, with its fluctuating climatic regime (see the sheep scab section below), then presumably the rodent flea (*X. cheopis*) was the main vector of the plague bacillus, as it is today in India and other parts of Asia? *X. cheopis* is a fur flea, meaning it primarily lives on its host during its life cycle, while the human flea (*Pulex irritans*) is a nest flea, living for the most part in the immediate surroundings of their human hosts. However, as *X. cheopis* is adapted for living in the fur of its host, it can and does adapt to the clothing of humans. Infected fleas readily kill their rat and other rodent hosts with the plague bacillus. It is thought that populations of black rats, moving in ships and other forms of transport with humans during the fourteenth century, were quickly killed off by their infected fleas and the fleas seeking new hosts readily bit the first available human that they came into contact with. Benedictow (2004) also points out another intriguing aspect of the rat flea biology. As grain is the primary food of the Black rat, which is why they are more commonly known as the house rat or ship rat, dislodged *X. cheopis* have developed the ability to live off grain and its debris for periods of time, although they require a blood meal to reproduce. Grain imports and the movement of grain from region to region may have been one of the primary mechanisms by which *X. cheopis* came into contact with humans, perhaps even independent of black rats (Benedictow 2004).

It is surprising, therefore, that *X. cheopis* has not been recovered from archaeological contexts in Europe during the Medieval Age, despite very favorable preservation conditions, particularly in deeply stratified organic deposits in towns. For example, *P. irritans* is frequently found in house floors, bedding areas and cesspits from the eighth to the fifteenth centuries in both urban and rural dwellings across Europe and the North Atlantic region (e.g., Buckland and Sadler 1989; Reilly 2003, 2012, 2018; Kenward et al. 2011; Smith 2012). Bird, cat, and dog fleas have also been identified. Morphologically, *X. cheopis* is distinct from these other species so it is unlikely to have been overlooked or misidentified. This has led some writers to speculate if, in fact, it was the human flea that spread the disease in parts of Europe, rather than the rat flea (Benedictow 2004; Eisen et al. 2006; Walløe 2008). Research in plague-infected areas of India, North Africa, and Madagascar, in the early part of the twentieth century suggested that the human flea was a poor vector of plague bacillus, even when houses are infested with them (Benedictow 2004). The argument goes that in order for the human flea to spread the disease it would have to ingest enough plague bacteria from an infected human so as to cause the "block" in its gut that characterizes the mode of disease spread by *X. cheopis*, i.e., the gut becomes blocked with bacteria and the flea regurgitates the infected blood when it next tries to feed, hence infecting the bitten rat or human. While this happens readily when *X. cheopis* bite rats and other small rodents, the numbers of bacteria in humans are considered to be too low to cause this blockage in human fleas (Benedictow 2004). However, other biologists disagree and argue that "unblocked fleas" may have carried sufficient bacteria to spread the disease, thereby allowing human fleas to play a role in early disease transmission (Eisen et al. 2006). It is argued that this might explain the rapid transmission of the disease from human to human in regions of Europe where *X. cheopis* was rare or absent.

So, it is possible that human fleas may have contributed to the spread of bubonic plague, perhaps particularly in severally immunocompromised individuals. The main culprit is still assumed to be the rat flea and perhaps other strains of the plague once it was unleashed, e.g., pneumonic plague and septicemic plague. There is some suggestion that lice also vectored the disease, although once again it is considered a poor vector compared to *X. cheopis*, while soil may also have been served as a temporary reservoir for the bacillus itself under favorable climatic conditions (Drancourt et al. 2006; Eisen et al. 2008; Hoffmann 2014; Campbell 2016). The absence of evidence in archaeological contexts for the rat flea, for the moment, cannot be considered to be evidence of absence.

Lice: Typhus in Medieval Europe and North America

Two main forms of typhus, epidemic and murine, are spread by insects, which act as reservoirs of the disease, though the latter form also uses small mammals, notably rats, as "amplifying" hosts (McLeod et al. 2010). In the context of the Medieval Age, evidence for outbreaks of the most virulent and devastating form, epidemic typhus, spread by body lice (*Pediculus humanus*), is the focus here. Zinsser (1935) presented the first detailed analysis of possible typhus outbreaks in Europe, starting with a typhus-like epidemic in Salerno, Italy, in 1083 CE. He proposed an eastern origin for the disease, possibly brought back into Europe by returning crusaders. He also proposed that the Plague of Athens, described by Thucycides in the fifth century BCE, was caused by typhus and was further evidence for its eastern or African source. However, the First Crusade did not begin until 1095 CE, some twelve years after the Salerno outbreak, and more recent research has suggested other diseases for the Plague of Athens, including typhoid fever (a different disease from typhus caused by the *Salmonella* Typhi bacteria) or viral hemorrhagic fever (Ebola or Marburg virus) (Dixon 1996; Olsen et al. 1996; Papogrigorakis et al. 2006, 2008).

Agger and Maschner (2009) suggest that the Rickettsia bacterium that causes epidemic typhus, *R. prowazekii*, was brought to Europe from North America as early as the eleventh century CE by Norse traders, via intermingling of human and animal lice in their North Atlantic colonies. Their theory suggests that Norse traders and their human lice became infected with the disease through contact with native flying squirrels, which carry the disease and may have nested in roof spaces in Norse sod houses over winter, and/or contact with human lice carried by Native Americans already infected with the disease. Returning along their Mediterranean trade routes to sell their goods in Italy, they brought the disease with them in their lice leading to the first outbreak in Medieval Europe in Salerno. According to contemporary sagas, Norwegians were involved as mercenaries in North Africa and Sicily between 1034–1041 (Page 1995; Hale 1998). In 1150 CE the Arab geographer al-Idrisi wrote from Sicily of the North Atlantic "inner island inhabitants" (Eskimos) and their use of large "sea animal" bones to construct their houses, clubs, darts, etc., presumably after hearing descriptions from returning Norse traders (McGhee 1984). Agger and Maschner (2009) speculate that the second recorded historic outbreak of typhus on the Italian peninsula in 1489 CE may also have arisen due to ongoing trade connections between the North Atlantic and Byzantium, rather than potential sources further east or south. Zinsser (1935) refers to writings by the Spanish conquistador Bernal Diaz recording an Aztec legend that described a typhus-like epidemic, which in 1116 CE led to the destruction of Tolan, capital city of the Toltec, the Mesoamerican state that preceded the Aztecs. Human lice have been found frequently on the scalps of mummified South American burials pre-dating the Spanish conquest and therefore the vector of epidemic typhus was certainly readily available in both South and North America (Araujo et al. 2000; Dittmar et al. 2003; Huchet 2015).

While different human parasites have been shown to have both New World and Old World origins, others, such as the bedbug *Cimex lecturlaris*, are considered post-Conquest introductions to the Americas from Europe (Ashford 2000; Bain 2004). The possibility that a disease such as epidemic typhus might present the opposite of this trend is intriguing, if difficult to prove definitively. Recent research in the genome of current strains of *R. prowazekii* suggests it is very closely related genetically to *R. typhi* (a global species), but more distantly related to other Old World Rickettsia (Raoult and Roux 1997). It is also related to the strain of *R. prowazekii* found in flying squirrels in America, although research has shown than different strains of the disease in humans are more closely related to each other than to those in flying squirrels (Bishop-Lilly et al. 2013). As pathogens are constantly evolving, however, this may not be surprising, and humans today can contract the form of typhus carried by flying squirrels (Bishop-Lilly et al. 2013). Therefore, the exact origin and timing of the first outbreaks of epidemic typhus in Medieval Europe is still an open question.

Mites: Sheep Scab in Thirteenth-Century England

The impact of (non-insect) arthropods, such as ticks and mites, on animal and human health in the Medieval Age is worthy of brief discussion. Slavin (2016) details the outbreak of scab in sheep in thirteenth-century manorial England. Animal disease epizootics are well attested in the historical record for the Medieval Age, notably the great cattle *panzootic* (i.e., animal disease outbreak over a large area, such as a continent) of the mid-fourteenth century (Newfield 2015a, 2015b). Through careful analysis of the records, it is assumed that most of these diseases took the form of direct animal to animal transmission, with rinderpest, which is transmitted by droplets from the cattle's nose and mouth, considered the most likely culprit (Newfield 2015b). The presence of an external vector or other causative agent is less easily identified in the record generally, with the possible exception of the eighth century CE equine epidemic outlined earlier (Gillmor 2005). This is why the sheep scab outbreak is worthy of note in the context of the theme of this chapter.

Scab is a highly infectious form of dermatitis caused by the feces and bites of the non-burrowing sheep mite (*Psoroptes ovis*). The mite has a two-week life cycle from egg to adult and the adults can live for up to forty days (Bates 2007). Importantly, mites can live away from their hosts for many days in wool sacks, woolen clothes, barns, and even soil during periods of colder weather (*c.* 10° C) (Slavin 2016). This latter characteristic of the mite may be part of the reason for its sudden appearance in the late thirteenth century, as this was a time of climate change across Europe (Büntgen et al. 2011; Hoffmann 2014; Campbell 2016). Ben-Ari et al. (2011) note, for example, that climate impacts all components of a plague/pandemic cycle—host, vector, pathogen—in various ways and over a wide range of scales. After a period dominated by warmer climates from roughly the ninth–twelfth centuries CE known as the Medieval Climatic Anomaly (MCA) or Medieval Warm Period (MWP), cooler wetter weather with longer and harder winters gradually set in across Europe, probably dating from the mid-thirteenth century (Hoffmann 2014). The causes may have been a combination of a series of volcanic eruptions during the thirteenth century, which increased sulphur aerosols in the atmosphere, as well as a long dip in solar activity, lasting from the 1250s to the 1380s, known as the "Wolf minimum" (Hoffmann 2014; Slavin 2016).

In the case of the sheep mite, this change to a cooler climate may have presented it with ideal breeding conditions. Slavin (2016) notes that the disease appears to have been largely confined to Britain and Ireland, where the years 1273, 1275, 1278, and 1280 were particularly wet and cold. The outbreak of 1279–1280 was the most notorious, decimating up to half the sheep stock of some English demesnes, while it was also reported in Wales and Ireland in 1280–1281 (Slavin 2015). It may have existed locally, causing small-scale outbreaks, in the years before the major epizootic event of 1279. The seasonality of the outbreak, with over half of all deaths occurring between October 1279 and February 1280, accords with the cold damp weather favored by the mite. This was also the time of year when sheep were corralled into stone or timber cotes during harsh weather, rubbing themselves off walls and undoubtedly encouraging the spread of scab. Slavin (2016) notes a new religious practice at the time designed to cure scab, which involved the gathering of sheep into cotes, the reading of Gospel passages, and sprinkling of holy water over the animals, but which may have inadvertently perpetuated the disease by these actions.

It is not clear if medieval sheep stock managers fully understood the cause of the disease, i.e., the presence of an external "aggressor" such as a mite, but the obvious physical manifestation of the disease (scratching, rubbing, loss of wool, skin infection) meant that various treatments were attempted during this time, including the application of pig lard, ox fat, grease, butter, oil, quicksilver, and tar (Slavin 2015). Of the manors reviewed by Slavin (2016) in the east and southeast of England, many recorded monies spent on "medicaments" for the treatment of scab.

DISCUSSION AND CONCLUSIONS

The case studies above give some sense of the impact of insects and other arthropods on the Medieval Age through their role as disease vectors. While we cannot be certain that epidemic typhus or Eastern Equine Encephalitis were the causes of the eighth-, eleventh-, and fourteenth-century outbreaks of disease in both humans and animals, we are on firmer ground with the thirteenth- and fourteenth-century outbreaks of bubonic plague and sheep scab and the various reports of malaria or malaria-type illness throughout the Medieval Age. All of these disease outbreaks took their toll, impacting population growth, agriculture, and economic growth for many decades in their wake, following on from the hardship wrought by famine and animal panzootics, particularly in the late thirteenth and early fourteenth centuries (Hoffmann 2014; Campbell 2016).

The link in the medieval mind between insect and disease, however, was not established at this time (Aberth 2016). Certainly, as we have seen, writers from this period understood the nuisance aspect of lice, fleas, and mosquitoes, and suggested various remedies for ridding oneself and one's home of them (Dalby 2011). The Hippocratic Corpus of texts, dating from the fourth–fifth centuries BCE, was still the main source of medical knowledge right into the Medieval Age. This meant that both lice and intestinal worms, for example, were seen as products of the human body, produced when the four "humors" of the body were imbalanced (e.g., blood, phlegm, black bile, and yellow bile) (Aberth 2016; Campbell 2016). The French surgeon Henri de Mondeville (d. 1320) believed that failure to expel "evil humidity" through sweating might lead to a catalog of skin diseases and that either fleas or lice (depending on the degree of heat on the skin's surface) could be generated as a result (Crawford 2007).

No link between, for example, flea bites and the plague were suggested in medieval literature. As Hoffmann (2014) notes, initial reaction to the Black Death was dramatic and panic-ridden. Jews, lepers, and other human scapegoats were assumed to have "poisoned" people and were persecuted. Apocalyptic movements, such as flagellants, developed and people speculated about natural forces and punishment by God.

Living conditions during the Medieval Age undoubtedly also contributed to the spread of disease. Mitchell (2015) notes that most people during this period did not have direct access within the home to clean water and privies/toilets. Pots, buckets, or other receptacles were used within the home for toileting and then emptied into a communal cesspit, yard midden, or town dung heap. One interesting story from an eighth-century Irish saga noted that it was not good to keep a "tub for washing or bathing" or a "slop pail" in the house after sleeping (Lucas 1965). The implication was that the content of the slop pail was human waste and should never be kept in the house at night.

Human feces were undoubtedly used as manure for outfields or garden plots within the town during the Medieval Age (Geraghty 1996; Mitchell 2015), ensuring intestinal parasites were "recycled" back into the human food chain. Oral-fecal contamination with intestinal parasites would, therefore, have been widespread and difficult to control contributing to general ill health and leaving the general population open to diseases from other sources, like biting insects (Crespo and Lawrenz 2014). Flies, in particular, were extremely prevalent during this time transferring pathogens from waste material to food on a constant basis. Fly species such as the house fly (*Musca domestica*), the stable fly (*Stomoxys calcitrans*), flesh flies ("blue bottles," *Calliphora* spp.) and the seaweed fly (*Thoracochaeta zosterae*) are often among the most numerous arthropod remains found in archaeological sites where ground conditions permit their preservation (Smith 2012, 2013; Reilly 2018). The latter, *T. zosterae*, is believed to have been attracted to the contents of cesspits during the Medieval Age and, although not necessarily a known vector of disease, may have inadvertently played this role through transfer of pathogens from cesspits to humans by alighting on food.

The understanding that animal and human waste material was dangerous to one's health was primarily bound up in the notion that noxious smells spread disease, rather than containing worms or other pathogens as the Irish Saga example above illustrates. Indeed, as Rawcliffe (1995) notes, this idea continued to find advocates into the eighteenth/nineteenth centuries. However, particularly in the wake of the Black Death, increased numbers of communal and private latrines were constructed within towns and rural communities during the fourteenth and fifteenth centuries (Mitchell 2015), while town "dung heaps" were moved further away from areas of dense occupation. At the same time, physicians were beginning to put together their own "plague tracts" based on careful observation of the disease and how it manifested itself and rejecting the classical "naturalist" views of where diseases came from. In this way, various quarantine measures and moves to clean up the urban environment were put in place (King and Henderson 2014). Local authorities sought to eliminate such sources of "miasmic" contagions as "… stynken caryn cast on the water nye the cytees or townes …," "… the corruption of privys …," and "… the casting of stynkyng waters and many other foule things in the stretes …" (Rawcliffe 1995, 2013; Hoffmann 2014).

While these measures were effective in improving general living conditions and (inadvertently) reducing the potential for disease spread by flies, they would have little impact on the diseases spread by insects such as lice and fleas. Indeed, along with mosquitoes, lice and fleas continue to play important roles in disease spread right up to the present day.

CHAPTER THREE

Insects and Food

SILVANA CHIESA AND GIOVANNI SOGARI

In-depth studies on the relationship between insects and food in the Middle Ages are generally lacking. Most scholars have examined this relationship in terms of protection of crops, but even this subject has been dealt with low relevance. The only exception of a strong link between insects and food is honey, an extremely important product for human gastronomy since antiquity.

MIDDLE AGES AND FOOD IDENTITY

The Middle Ages or Medieval Period, which lasted conventionally from the fifth to the fifteenth centuries, is a period specifically related with European history: the collapse of the Roman Empire, the spread of Christianity, the meeting and the conflict of different cultures (the Roman and the "Barbarian"), concluding with the birth of Europe and the definition of its religious identity. The building of this new (Christian) identity involved every part of human life and culture, including a particularly complex definition of food identity.

Even though the diet of the Romans and the Barbarians included more or less the same foods (meat, cereals, cheese, etc), their place, importance, and role were different in each nutritional system (Montanari 1994). On the one hand, the Romans and the Mediterranean people had represented themselves as farmers, people able to "create" their own food, and as Massimo Montanari wrote in *The Culture of Food* (1994), "... they developed a diet, which we may call 'Mediterranean', characterized by a dominant vegetable component (grain preparations and bread, wine, olive oil and greens) complemented by a little meat and goat cheese" (Montanari 1994: 6). On the other hand, "... Meat rather than bread, was the most important element of their (the Barbarians) diet. Instead of wine they drank milk or cider (made by fermenting wild fruits) or beer ... Instead of oil, butter or lard was used for greasing or cooking" (Montanari 1994).

According to Flandrin et al. (1999), the rise of Christian religion led to the fusion, or, more accurately, the acculturation of both food cultures: every kind of food was permitted, but not at every time or every day. There were some days when eating meat was allowed, and some on which it was prohibited, like a sort of penitence. Attention was focused on the dichotomy between food that was thought to be "good for the body" and food that was "good for the spirit." Everyone considered the body and its necessities (food, sex, and pleasure) a limit for reaching the state of *holy person*, and everyone, in the first centuries of the Medieval Ages, wanted to become a holy person: a concept which became the basis of what was called a "monastic food and dietetic," which was based primarily on vegetables, cheese, and, extremely rarely, meat.

Starting from the sixth–seventh centuries, Europe was considered completely Christianized, and this fact might be the main element of a common identity and the reason why food behaviors were similar in every region or state. Christianity does not consider any food prohibited, a fact in stark contrast to Judaism (and Islam later) where the food regime was based on the rules in the Holy Bible (Torah and Talmud). The absence of forbidden foods, however, does not mean that there were no food rules; on the contrary, there were a lot of procedures for preparing, cooking, and serving food and all this modus operandi was collected in dietetics and cooking books.

Having discussed the background, the rest of this chapter will investigate the relationship between food and insects during the Middle Ages.

Although we are conscious that food culture and food behaviors are nearly always practical knowledge and rarely contained in written books, the written sources are the only ones possible to study. Archaeology might make available some information about food behaviors in the past, but, unfortunately, on the specific topic of the insects we have not found any trace or report in archaeological literature related to the Middle Ages. Despite the fact that books were rare before the invention of the printing press (around the end of fifteenth century), several handbooks existed that focused specifically on animals, food production, and human health. These manuals provide us with the opportunity to explore the presence of insects and their relationship with food in the medieval times. To simplify the study, these sources are grouped into naturalistic books, agronomy texts, dietetic and cooking books, and narrative and religious sources, including penitential books.

Modern and contemporary sources focusing on insects and food, such as *Insects as Food: History, Culture, and Modern Use Around the World* (Costa-Neto and Dunke 2016) or *Insects as Human Food: A Chapter of the Ecology of Man* (Bodenheimer 1951) do not contain much useful information on this subject. Some interesting information, which has been taken into consideration in this chapter, can be found in *Crop Protection in Medieval Agriculture: Studies in Pre-Modern Organic Agriculture* (Zadoks 2013).

NATURALISTIC BOOKS

During the Middle Ages and until the fourteenth century, two very influential works dominated the field of zoology in Europe: the *Historia Animalium* (*History of Animals*) written by Aristotle (4 BCE), and the *Naturalis Historia* (*The Natural History*) by Pliny the Elder (79 CE). Ancient Greek philosopher, naturalist, and scientist Aristotle (338–322 BCE) can be considered the father of the science of classification and taxonomy of living organisms, in particular animals and plants. Aristotle first classified animals and plants and grouped them by their similarities: animals with blood or not, living in water, on land, or in the air. The classification system proposed by Aristotle had obvious limitations; for example, birds, bats, and flying insects were all grouped together, even though they have little in common except the fact that they can fly (Balme and Peck 2014). The Roman natural philosopher Pliny the Elder (23–79 CE) continued this tradition in his *Naturalis Historia*. This work is divided into thirty-seven books organized into ten volumes, with the eleventh book focusing on the insects as Pliny understood them (Maspero 2011).

One of the reasons these two books became so famous was that such a classification system provided a hierarchical order of importance that placed Man at the top and the insects at the bottom, a vision that perfectly reflected the medieval idea of a world created by God. The Christian culture used these works to reinforce the theological view of nature, and animals became but a "metaphor" for human qualities and defects. This also applied to the insects: ants, for example, represented a perfect organized society and hard collective work. The bees were better than the ants because of honey and wax production. Locusts were a symbol of devastation as they are associated with the plagues of Egypt (Figure 3.1). This is one of the earliest documented connections between insects and food, and reflects the fear that people had when cereals (and in general food) were lost because of insect attacks.

Pliny the Elder not only classified insects from an anatomical point of view, but attempted to describe the advantages and the disadvantages they can provide to humans. He focused particularly on bees, wasps, hornets, bumblebees, cicadas, beetles, locusts, ants, moths, and mosquitoes, as well as spiders and scorpions, which are notably not insects. Thousands of copies of the *Naturalis Historia* were reproduced and, acquiring commentaries, they spread to every library in castles, monasteries, abbeys, universities, and in every bookshop in

FIGURE 3.1 Locust swarm, medieval Bible. Credit: PRISMA ARCHIVO/Alamy Stock Photo.

Europe from the sixth to the fifteenth centuries (Maraglino 2012). Despite the fact that it was written more than five hundred years earlier, this book remained the only taxonomic text in Europe for centuries.

The *Bestiary*, a compendium of real or fictitious beasts, became a very popular literary genre in the Middle Ages. The *Bestiaries* were considered the descendants of the *Naturalis Historia* since they contained accurate descriptions of the animals embellished by tinted illustrations. These books were treated as educational books, where every species of animal was the pretext to justify some moral metaphor. In this framework, the presence of insects was limited to only a few: the ants, because they were thought as the paradigm of foresight; the bees for their social organization and for the production of honey and wax; and the spiders for their wisdom and the ingenuity in stretching the threads of their canvas, waiting for the fly to fall into their trap. Sometimes the cicadas and the flies were also mentioned because of their negative reputation: cicadas are always singing and never working, and flies are annoying, always buzzing around. In these books every insect was always either a positive or negative example for the reader, but they were never linked with food.

AGRONOMY TEXTS

The most important agronomic book in the Middle Ages was the *Liber cultus ruris*, better known as *Liber ruralium commodorum* (Book of Rural Benefits) or *Ruralia commoda* by Pietro de Crescenzi (1235–1320). It was written sometime between 1304 and 1309 in Latin but was quickly translated into Italian, and by the end of the fourteenth century (1373), it was also translated into French for King Charles V. Like most of the medieval works, *Liber of ruralium commodorum* was based on previous classical books and was even structured in twelve parts like its predecessors, the *De re rustica* (*On Agriculture*) by Lucius Junius Moderatus Columella (first century CE) and the *Opus agriculturae* (*Work on Agriculture*) by Rutilius Taurus Aemilianus Palladius, on which most of de Crescenzi's book was based.

The agronomic manuals focused on the best ways to manage a farm and to produce food: from the cereals, vegetables, and fruits all the way to obtaining meat from animals. These manuals were compiled like a comprehensive catalog of all the foods known in the Middle Ages. Insects were not included in the list of foods. On the other hand, insects were considered and mentioned as a cause of food losses, damage, and contamination of both ingredients and in processed foods; for example, weevils were noted as common insect pests in wheat and flours (Figure 3.2). One of the most common pests is the so-called wheat weevil (*Sitophilus granarius*), an abundant grain pest encountered in many archaeological deposits (Smith and Kenward 2011). One reason for its abundance might be explained by the difficulty to sieve out weevils, due to their larger size which is closer to the grains; moreover, the females infest the grain by laying their eggs inside of it, which then become larvae (King et al. 2014). The infested grains found in archaeological sites may not have always been intended for human consumption but perhaps for animal feed. The two other major coleopteran grain pest species were *Oryzaephilus surinamensis* (L.) and *Laemophloeus ferrugineus* (Steph.).

Thanks to archaeological studies, we also know that the pests of stored grain also spread throughout history. For instance, large amounts of grain in the Roman period was destroyed by these insects (Smith and Kenward 2011). Moreover, it seems that before the arrival of the Romans there were no records of grain pests in Britain, which suggests that with the spreading of the Roman Empire, the wheat weevil was introduced into a non-native environment (Buckland 1981).

Piero de Crescenzi paid special attention to insects with respect to the precautions that the farmers had to take in order to avoid possible injuries from the growth or conservation of cereals, vegetables, or fruit trees. In chapter 3 of *Liber of ruralium commodorum*, he wrote: "the grain must be protected by covering the grains harvested with fresh olive leaves to prevent the 'gorgoglioni'

FIGURE 3.2 Granary weevils (*Sitophilus granaries*). Credit: Wikimedia Commons.

(possibly a weed beetle) damaging them." Beetles are also considered the most important group of insects infesting grains by King et al. (2014) and Smith and Kenward (2011), who focused on the archaeological presence or absence of grain pests in Britain from the roman period to the Middle Ages.

Unfortunately, Piero de Crescenzi did not study this subject in-depth and these few passages are all the information we have. But he again refers to insects in his fifth chapter devoted to different fruit-crops to suggest how to avoid attacks by "worms," a term that included all kinds of parasites. The trees mentioned are fig, plum, peach, as well as mulberry (for its leaves being the food for the precious silkworms) and laurel for its symbolic value.

The invention of printing press in mid-fifteenth century resulted in massive diffusion of printed books in the Late Medieval Period, among which were several treatises on agronomy. Agostino Gallo (1479–1570) was an Italian agronomist best known for his *Le vinti giornate dell'agricolatura e I piaceri della villa* (*The twenty days of agriculture and the pleasure of villa*). The first edition of this book, which was published in 1564, was titled *The ten days of agriculture* ..., but a few years later it became *thirteen days* and finally *the twenty days of agriculture*. This book describes, day by day, all the work needed to be done on a farm, suggesting new techniques and best practices to solve every problem related to pests. From Pliny the Elder to Agostino Gallo, agriculture manuals became more and more specific and scientific, but the nomenclature for pests and particularly the insects remained poor and general: most of them were called "worms." There were good worms and bad worms, but the reader often could not distinguish these from one another because very little attention was paid to proper morphological descriptions.

In the first chapter (or first day), Gallo (1569) directed his attention to a particular kind of "bad worm" which were thought to be very dangerous for the roots of the wheat, which he calls *zaccarole* (a term not found in any dictionaries). The author described how to discover their presence in the ground, how to collect them, and finally suggested giving these worms as food to the hens. He goes on to say that these worms were "extremely cold in nature" because they came out when the ground was warm. He recognized that the worms were edible, but *not* for Christians, only for animals, particularly for the hens—which until modern times were thought to consume worms as their natural food preference.

DIETETIC AND COOKING BOOKS

During the ten centuries of the Middle Ages, dietetic books were strictly concerned with cooking, since the food regime was considered in the realm of medical knowledge and cooking prowess. According to Capatti and Montanari (2003), "The basic principle was that every living thing (human, animal, vegetable) possessed a particular nature based on the combination of four qualities," which in turn were considered antithetical to one another: hot, cold, wet, and dry. Each type of food was classified according to the presence, prevalence, or absence of these qualities. The consequence was that if a vegetable was considered too wet (e.g., the onion), it had to be fried or roasted or combined with a dry food. Animals' meat was classified the same way: "Venison is eaten boiled; roasting is good if the deer is young" (Montanari 1994). It is evident that cooking and dietetic books strived to be full of exact indications of the "quality" of every kind of food. In this regard, if something was considered food, it had to be included in a dietetic or in a cooking book, however there are no traces of insects in medieval dietetic books with the exception of a mention for being careful to avoiding the presence of larvae in flour and rice.

One of the very few connections between insects and food in medieval times was the attention people paid to honey. Honey was very popular and considered for its many benefits. Paragraphs specifically dedicated to honey can be found in many dietetic books, including *De honesta voluptate et valitudine* (*About the honest pleasure and the good health*) (1474) written by Bartolomeo Platina, in which bees are revered as "neat" insects in contrast to other "not neat" insects, such as flies. Platina also observed differences in smell and the taste of honey related to the quality of plants and flowers that the bees met, and underlined that, in his opinion, the best honey was from Sicily and North Africa because of the presence in these areas of a huge quantity of thyme (aromatic herb). The worst honey, he surmised, was that which came from Spain and Pontus (the Black Sea region of Turkey).

From a dietetic point of view, honey was classified as "hot" in nature, and for this reason it was thought to alleviate numerous maladies (particularly those caused by cold). Honey collected during the summertime was considered the best, and it was considered better cooked than raw; it was used in preserving apples, pumpkins, citrons, cedars, and walnuts. Honey was widely used in the kitchen, also to prepare the *Idromele* (mead): "which can be called honey water" useful against coughing (Platina 1474). Mulsum was made by adding water and honey to wine. Mead was a popular alcoholic drink in the Middle Ages; it was made by fermenting the final washings of honey from the comb in a solution with water (Allsop and Miller 1996).

NARRATIVE AND RELIGIOUS SOURCES AND PENITENTIAL BOOKS

Using religious narratives and penitential books can be very difficult from a historical point of view. On one hand, the sources rarely report the events accurately, and often utilize hyperbole or metaphor to describe what happened. On the other hand, these narratives reflect the mentality of the period which makes them worthy of a mention here. Some examples may be the chronicles of the terrible famine and starvation that occurred by the end of the fifth century, and, more frequently, during the thirteenth and fourteenth centuries in Europe (Montanari 1994). In these chronicles, people are sometimes described as eating everything out of hunger and desperation. The list of unclean food allegedly consumed is terrible: meat from corpses, soups made with rats or insects, bread made with grape seeds or hazelnut blossoms, and fern roots pressed dried and powdered. Other instances include episodes of extreme hunger or desperation, such as hermits who lived in the deserts, or prisoners deprived of food. Regardless of whether or not these stories were true, they lead us to the conclusion that entomophagy was viewed as an "extreme" eating behavior, tolerable only before the death.

The approach of the medieval religious thinkers, writers, and philosophers to the description of John the Baptist eating locusts in the Bible is rather interesting: "John wore a garment of camel's hair and a leather belt around his waist and his food was locusts and wild honey" (Mark 1:6; Matthew 3:4). On one hand, apparently the use of the specific adjective "wild" relating to honey suggests that the situation was not normal and everything should have been on the border between civil and uncivil, or domestic and wild. Consequently, if the honey was wild, then eating locusts as food should have been tolerable: not normal, but acceptable, understandable, and proper for an extreme situation. On the other hand, the philosophical debate shifted from eating locusts (whether acceptable or not) to defining whether the locusts should be considered as meat or not meat. St. Jerome (393) in his famous work *Adversus Jovianum* (Against Joviannus) wrote that in the desert, John the Baptist ate locusts, not meat (Migne 1845). On the contrary, St. Augustine of Hippo wrote that people were allowed to eat meat because John the Baptist ate locusts in the desert and he did not sin (Dods 1950; Ryan 1960). Apparently they were not shocked by the mere fact of humans consuming locusts, probably because the more pressing issue during the Middle Ages was the acceptability (or lack thereof) of eating meat by the religious men, particularly monks and priests. The debate was so pervasive and discriminating that people's attention was directed towards defining if locusts were meat or not. Indeed, St. Jerome correctly pointed out that there was no single food common to all people, because each human group eats what is better and most abundant in their own land. He argued that there were many groups of people who ate only fish, and others who ate *worms* (Montanari 1999). The fact that most people belong to niche cultures is anthropologically true and relevant even today.

During the Middle Ages, due to the religious wars (the Crusades) and frequent episodes of Jewish persecution, eating certain foods or not eating some others became an "identity food" phenomenon: offering pork meat to someone was considered a practical way to discover whether that person was Muslim or Jewish, because this kind of meat is a taboo for them. Christianity, however, did not have any forbidden foods, and therefore it is surprising that there seems to be no traces of entomophagy is Christian Europe. Penitential books, manuals used by priests of the western church especially during the Early Middle Ages to administer ecclesiastical penance, offer clueş on whether or not eating insects was considered a censored behavior: a sin. These books contain detailed lists of sins that the priest was to consider in assisting an individual penitent, with his examination of conscience and confession during the rite and corresponding penances or acts that were to be assigned to the penitent (Encyclopaedia Britannica 2023). In Penitential of Burchard, bishop of Worms, it is established that "if a person eats lice (voluntary or not), the penitence will be at least seven days of nothing but water and bread." Seven days was a very short and light penance, since the penances usually began from twenty days of water and bread. The reason may have been that grooming for picking out lice in a family group is an anthropologically well-known behavior, and inadvertent eating of lice was probably a quite frequent behavior; not appreciated, but also hardly rebuked.

THE CASE OF THE LOCUSTS AND CICADAS

The relationship that scholars of the Middle Ages had with locusts and cicadas is strange, perhaps because they had not yet understood how the calm cicadas, who lived in leafy trees such as lindens and spent all summer "singing," would turn into fierce devastators of fields and become the cause of famines, misery, and death.

As mentioned before, the most studied, printed, and copied book of the Middle Ages was *Naturalis Historia* by Pliny the Elder, who divided cicadas into two groups: the smaller ones (tettagoniae) and those who sing (achetes):

> ... The cicadas also live in a similar manner, and are divided into two kinds. The smaller kind are born the first and die the last, and are without a voice. The others are of the flying kind, and have a note; there are two sorts, those known as *acheta*, and the smaller ones called *tettigonia*: these last have the loudest voice. In both of these last-mentioned kinds, it is the male that sings, while the female is silent.[1]

He also mentions that other authors distinguish between those who live and feed on the shoots of ash trees, and those found on wheat and/or ripe oats: "... Some authors divide these animals into two kinds, the *surcularia* which is the largest, and the *frumentaria* by many known as the *avenaria*. This last makes its appearance just as the corn is turning dry in the ear."[2]

Pliny later writes about locusts, and specifies that they have strong hind legs that are folded towards the outside. He mentions that locusts were considered a scourge sent by divine wrath, reaching Italy from Africa, traveling enormous distances, covering farmed fields like a menacing cloud, and devouring everything one bite at a time. He tells of the story that in Cyrenaica (in modern-day Libya), in order to fight locusts, there was a law that required destroying them (literally to "wage war") three times a year: first to destroy their eggs, then the young, and finally the adults. Sparrows of the jay species were specifically bred to fight them, and in Syria even the army was used against them:

> This is looked upon as a plague inflicted by the anger of the gods; for as they fly they appear to be larger than they really are, while they make such a loud noise with their wings, that they might be readily supposed to be winged creatures of quite another species. Their numbers, too, are so vast, that they quite darken the sun; while the people below are anxiously following them with the eye, to see if they are about to make a descent, and so cover their lands. After all, they have the requisite energies for their flight; and, as though it had been but a trifling matter to pass over the seas, they cross immense tracts of country, and cover them in clouds which bode destruction

> to the harvests. Scorching numerous objects by their very contact, they eat away everything with their teeth. Those from Africa are the ones which chiefly devastate Italy ... In the territory of Cyrenaica there is a law, which even compels the people to make war, three times a year, against the locusts, first, by crushing their eggs, next by killing the young, and last of all by killing those of full growth; and he who fails to do so, incurs the penalty of being treated as a deserter ... It is for this reason, too, that they breed the jays, who fly to meet the locusts, and kill them in great numbers. In Syria, also, the people are placed under martial law, and compelled to kill them: in so many countries does this dreadful pest prevail.[3]

In the books from antiquity it is not always possible to make any net distinction between "*cavallette*" (grasshoppers), "*locuste*" (locusts), and "*grilli*" (crickets) in terms of modern systematic entomology (Camuffo and Enzi 1991). It is not surprising that all of these expressions were sometimes considered synonyms. For instance, Matteo Villani, an Italian historian and writer (Florence 1283–1363) systematically used the term "crickets" instead of grasshoppers, while other authors describe the same invasions using the term "locust" (e.g., Schiavina 1616, *Chronicum Parvum Ripaltae*, 1363). Today we know that grasshoppers, locusts, and cicadas are Insects belonging to the order *Orthoptera*, stemming from the ancient Greek and meaning "straight wings." The group is further divided into *Ensifera* to which the *Tettigoniidea* families (green grasshoppers) belong, and *Caelifera*, to which the *Tetrigidea* (dwarf grasshopper) and *Acrididea* (locusts) families belong.

Grasshoppers can exist in solitary form or small groups, and in large groups (swarms). Today we know that in practice, when the weather becomes too hot or otherwise unfavorable, and the food becomes scarce, especially when the density of insects becomes excessive, that is when the grasshoppers "touch" one another, which leads to the release of a neurotransmitter, *serotonin*, triggering a series of biological and morphological changes that initiates the "herding" process: grasshoppers turn into locusts and gather in swarms that travel for many kilometers looking for food (Anstey et al. 2009). As reported by Hatcher and Battey (2011), locusts are defined by their behavior of changing their status from a solitary individual phase to a gregarious one, becoming a real "swarm," with a density of 30–150 individuals per square meter.

In this transformed phase, the population of locusts can became a serious plague. Van Huis et al. (2007: 193) define plague as an event that occurs "when widespread infestations of swarms and hopper bands affect extensive areas and generate large numbers of reports during the same year and in each of successive years." This epidemic is different from the presence of many grasshoppers, which even if they are considered as an agricultural pest, there is no formation of a swarm population. The transformation between the solitary and gregarious phases of the locust is complex and triggered by the presence of a high density of other individuals. This phase-change usually happens in a matter of a few hours, and the mechanism involves behavioral, morphological, and physiological changes (Hatcher and Battey 2011). Weather conditions (e.g., rainfall or winds driving individuals into the same area), structure, and quality of vegetation and other factors allow reproduction.

When locusts are not swarming, they need to stock fat reserve in their body, consuming a large amount of food. For example, an adult desert locust may consume in a day around 1.5 grams of vegetable, and a swarm can eat a hundred to a thousand times as much plant material in a pasture as livestock would (Hatcher and Battey 2011). The formation and movements of these swarms have remained unexplained and feared for centuries, which is why, since ancient times, people used to think that it was divine "punishment." While the biology of grasshoppers/locusts was not well known during the Middle Ages, the damage they caused was; indeed, often these damages were so serious that individual episodes thereof led to narrative exaggerations that were passed on and uncritically repeated. Typically, this affected the number of deaths resulting from the locust invasions, which sometimes was three hundred thousand, sometimes eight hundred thousand, and sometimes the whole country.

St. Augustine is also involved in these narratives. In his *De Civitate Dei*, he writes:[4]

> *They wrote* that the number of grasshoppers in Africa, when it was a Roman province, was prodigious; *they say* that, after destroying the fruits and the vegetation, they threw themselves into the sea like a cloud so huge to be beyond any calculation. There they died and were returned to the beach. As the air became infected, a serious epidemic erupted and, as it is told, eight hundred thousand individuals died in the kingdom of Massinissa alone and many more did in the regions near the sea.[5]

We would like to emphasize the questionable historical reliability of a narration based on *dicunt*, which sounds more like a "popular rumor being told" rather than a reference to one or more accurate historical subjects.

Still on the subject of popular beliefs, it seems that St. Augustine was even personally involved in what is called the miracle of the locusts in Toledo: during the year 1268, Augustine was prayed to by the inhabitants of Toledo to rid them of the disasters caused by the locusts, which not only threatened to destroy the crops and starve the town, but also to fill it with a nasty plague, because after eating, [the locusts] died, and by decomposing quickly they left an intolerable stench, which was a clear indication that the noble town of Toledo was soon to be hit by the plague (Figure 3.3). For this reason, the inhabitants of Toledo decided to call for the heavenly help of the Saint, who appeared to the people of Toledo and "by virtue of Heaven, he made it so that all those ugly animals rushed themselves into the nearby river Tagus." Several testimonies are given of this miracle (Alvarez 1959), but above all, the miracle and devotion to St. Augustine by the inhabitants of Toledo becomes the subject of many iconographic representations:

- Fanzago Cosimo (1591–1678), Certosa di San Martino Museum, Stalli dalla chiesa di S. Agostino, Naples, Italy
- Portrait of Schelte à Bolswert (1624), printed edition, Paris, France
- Malpiedi Domenico (1630–1640), Sant'Agostino Convent, San Ginesio, Italy
- Begni Giulio Cesare (1640), S. Agostino Convent, cloistre, Fano, Italy
- Miguel de Santiago (1656), S. Agostino monastery, Quito, Ecuador
- Juárez Juan Rodríguez (1730), Picture gallery of San Juan de Sahagun convent, Salamanca, Spain
- Melendez Miguel (1734), Prado Museum, Madrid, Spain
- Melendez Miguel (1734), Museum of fine arts, Segovia, Spain

Going back to the descriptions of locust invasions, a peculiar detail is that many medieval chronicles, both from the Mediterranean and the Spanish regions, report very serious famines and pestilences as a consequence of the devastation of the fields. While famines are widely understood as the loss of

FIGURE 3.3 St. Augustine vanquishing the Plague of Locusts, by Miguel Jacinto Meléndez (1734). © Museo Nacional del Prado.

crops and the destruction of produce and pastures, the "pestilences" were almost always attributed by chroniclers to the putrefaction and nauseating reek coming from the insects' corpses, as they died en masse and infected the air (Segni 1605). Following the plague epidemic that struck and devastated Europe in 1347–1348, any episode involving collective deaths was deemed suspicious, and when taking into account the biblical suggestions, the general weakening of the population which was certainly more exposed to infections and diseases due to the lack of food, and the frightening phenomenon of darkening of the sky by the swarms of locusts, it reminded people of the arrival of the three knights of the Apocalypse: Hunger, War, and Plague (Aldrovandi 1602).

Starting around the mid-fourteenth century, locust invasions became more frequent both in the Mediterranean region (southern Italy, Sicily, Malta, Cyprus, Greece, Tunisia, Algeria, Morocco, etc.) and in Europe (northern Italy, Spain, Germany, Hungary, Croatia). These events were most frequent during the summer, especially August, and were frequently associated with a real drought. The cause of such epidemics was probably due to several reasons: the invasion of insects, followed by the physical weakness and ill-health of the population, as well as political and social difficulties (e.g., commerce and transport of foodstuffs).

An interesting work of research by Elina Gugliuzzo and Giuseppe Restifo (2014) follows up on the most famous and devastating locust invasions in Europe and Italy from the Middle Ages to the end of the nineteenth century. The authors mention the studies already published by Richard Hoffmann (2010) and others, reporting the devastation in Spain in the periods between 1270 and 1360 and 1530 and 1590. The consequences of these invasions were so important that even the well-known Italian naturalist Ulysse Aldrovandi dramatically listed them in his *De animalibus insectis libri septem, cum singulorum iconibus ad vivum expressis* (1602): in 1475 in Hungary, Moravia (the eastern part of the Czech Republic) and northern Italy; 1477 in the countryside near Brescia and Mantua; 1535 in Poland; 1541 in Poland and Romania; 1542 in virtually all of Europe: Italy, Germany, Silesia (Poland), Saxony, the Adriatic region; and then again 1564–1566 several times in Italy and 1565 in Cyprus. Gugliuzzo and Restifo report a very serious episode in 1355 in Sicily, in 1363 in Sicily and then again in Puglia, Marche, and Romagna, and in 1478 in Friuli.

Since ancient times, it was suspected that locusts travel through the sea pushed forth by the winds. In the Middle Ages this idea grew stronger, and today the importance of the winds in defining the intensity and direction of the locust invasions is widely confirmed (Gugliuzzo and Restifo 2014). After the sixteenth century, the epidemics of famine significantly decreased thanks to several factors: first, growing technological developments, better hygiene, and increased commercial exchanges over a broader territory. In addition, after the discovery of America in 1492, more crops (e.g., grains) were cultivated, assuring a more diversified variety of food (Camuffo and Enzi 1991).

CONCLUSION

This chapter has described the relationship between insects and food in medieval times. Our investigation, through the study of well-known agronomic, dietetic, and naturalistic books written before and/or during that period, suggests that insects are never linked or listed as food. The only connection is the importance that people have given to honey, an important commodity and a popular food item. Even if the presence of citations and mentions in written books are rare, we know how these little animals have been crucial for the evolution of human and eating behavior, considering their role as pests in destroying stored grains (e.g., wheat weevils) and as destructive insects in damaging crops (e.g. locusts when they become swarms).

We know that the definition of food identity based mainly on religion (Christianity) brought people to eat mainly vegetables, cheese, and, less frequently, meat. However, insects were not considered meat and the fact of eating them was considered a censored behavior, as reported in the penitential books.

CHAPTER FOUR

Insect Products

GENE KRITSKY

The Medieval Age, also called the Middle Ages, is the historical period between antiquity and the Renaissance, and lasted approximately a thousand years. It began with the fall of the western Roman Empire, when Flavius Odoacer deposed Emperor Romulus Augustus in 476 CE to become the first King of Italy. The eastern Roman Empire, also known as the Byzantine Empire, was centered in Constantinople, which endured for nearly another thousand years before falling to the Ottomans in 1453.

The Medieval Age is often divided into three periods. The Early Medieval Age, also called the Dark Ages, encompassed the first five hundred years. The Dark Ages are aptly named for the lack of written records during this period, and this lack of documentation is reflected in our appreciation of insect products used during this time. Village life in the year 500 was not that different half a millennium later at the start of the High Medieval Age, which includes the next three centuries ending around 1300. It was a period of increasing population in Europe especially in the urban centers, and our knowledge benefits from a greater number of written records. The Late Medieval Age that followed was marked by famine and the Black Death, and ended around the year 1500.

The start of the Medieval Age is not marked with significant technological changes, and this is clearly evident when we examine insect products from honey bees. In Egypt, for example, beekeepers continued to keep bees in long tubes made of mud and straw that were stacked into apiary walls as they had since the construction of the Great Pyramid. In Greece, pottery hives with closed ends were still in vogue while pottery tubes were in still used in the Iberian Peninsula (Kritsky 2017).

There is very little information involving beekeeping during the Early Medieval Age. In fact, the hives illustrated in the Twenty-Sixth Egyptian Dynasty tombs of Pabasa and Ankhhor will be the only beehive illustrations for approximately 1,100 years. We can confirm that beekeeping continued to be an important occupation in the Early Medieval Age as there are records that mention honey (Kritsky 2017).

The kinds of hives in use at the start of the Medieval Age are thought to be similar to those in antiquity. Hives were made from hollow logs, fashioned out of cork, wooden fennel stalks, pottery, and even woven out of wicker. The only physical evidence of beekeeping dates just before the start of the Medieval Age and includes two log hives dating between 200 and 500 CE from northern Germany. The first woven wicker skep hive, which dates to 200 CE, was found in northern Germany, and it included part of the crown and the bottom board, with the crown portion verifying that it was round.

Woven wicker skeps are distinctly conical shaped with a more triangular appearance. This shape is the result of their construction that starts with a cylindrical or conical piece of wood called the crownpiece that serves as the anchor for the wooden vertical ribs or "warps." These warps are connected to the top by wrapping the ends of the ribs with a strong grass or inserting the ribs' ends into holes drilled into the crownpiece. The "weft," the material that was woven around the "warps," was first soaked in water to make it more flexible. The warps may have extended beyond the rim of the base raising it up an inch or two to serve as an entrance for the bees. Two sticks would be placed inside the skep about 15 cm from the top to provide the bees with an internal scaffold onto which they could start to build comb. To protect the skep from the elements, the outside would be coated with cow dung or mud and allowed to dry (Kritsky 2010; Speis 2016).

Illuminated medieval manuscripts often include images of beehives in use during the High Medieval Age. It is not always clear what materials were used to construct the hives except for skep hives, which were either made of wicker or straw. Generally, illuminated manuscript skeps that include a crownpiece or distinct ribs are believed to be made of wicker whereas skeps without a crownpiece and sometimes fashioned with a small rounded handle at the top are thought to be made of straw. A thirteenth-century Bestiary at the Bodleian Library at Oxford includes an illustration of a wicker hive with a top or crownpiece. Another thirteenth-century Bestiary in the Bibliothèque Nationale de France includes an illustration of a mud-covered wicker skep. Finally, a Late Medieval Age manuscript includes a detailed image that illustrates the construction of the wicker hive with the ribs radiating from the crownpiece (Platearius n.d.; Barber 1992; Bodleian 2016).

Straw skeps (Figure 4.1) are made from coiled straw that is stitched together with briar or cane. The thickness of the straw coil is controlled by a girth

FIGURE 4.1 A wicker hive (left) and a straw skep hive (right). Note how the wicker skep is made out thicker plant material that is woven between vertical ribs. The straw skep is made from coiled straw that is tightly stitched together with briar. Photograph by Gene Kritsky.

made from horn. Coiling or twisting the straw increases the strength of the straw in much the same way that coiled rope is strong. The skep starts with a tight whorl. The shape and size of the skep is determined by how the coils are stitched together (Kritsky 2010).

Straw skeps had many advantages over wicker skeps. Hives made of straw were lighter than those made of wicker, and the coils of straw could be stitched tighter together than wicker, which encouraged the bees to coat the interior with propolis (plant resins the bees collected). This coating provided greater protection against inclement weather. Wicker skeps, on the other hand, needed to be covered with dried mud and or dung to keep out rain, but this covering was prone to breaking off exposing the bees inside. A good straw skep was quite durable and could be used for over a century (Kritsky 2010).

It is thought that straw skeps originated in the Elbe River region in northern Germany. The oldest illustration of a skep dates back to the late 1400s, but straw skeps certainly predate the Late Medieval Age. According to the *Oxford English Dictionary*, the word "skep" has its roots in the Old Norse word "skeppa," which has roots to the Old Saxon "scepil." "Skep" dates back to 1100 when it described a unit of measure as the amount of coal or grain that a specific sized basket could hold. That amount was a "skepful." By 1300 the English word skep was used for a strong basket that could hold coal. The oldest word use of "skeps" with bees is dated 1494 in a deed that read "The same Kateryne shall have fre yssve to goo and come to hyr hyve skeppys beyng [within] the Meese and Yards." The written use of skep as a hive was ninety years later (Kritsky 2010).

These dates are upper limits for the use of the word "skep." The OED notation does not suggest that the word and the hive were created at the same time. Rather it documents when the term was first used in a written document. The paucity of written records during the Early Medieval Age is such that the word for skeps was likely used centuries before the word "skep" was used in a written document (Kritsky 2010).

The use of both wicker and straw skep hives, both made from perishable materials, required that they be protected, and many of the illuminated manuscripts document how this was done. Several of the hives were placed on stools or small tables to lift the hives off the ground. Others show the hives on ledges or shelves on castle walls, whereas a few show the hives placed under lean-tos (Kritsky 2017).

Permanent protection in the form of special bee niches were constructed in castle, abbey, garden, and building walls (Figure 4.2). A few medieval bee niches, also known as bee boles or bee garths, can still be found in the United Kingdom. Five survive from the High Medieval Age with the oldest dating back to 1250 and twenty-one documented niches remain from the Late Medieval Age. These were made of stone or bricks. Some were simple square sided niches, whereas others were more elaborately built with rounded backs with a rounded top or a more gabled shaped top (Kritsky 2010).

Skeps were not the only hives to be employed during the Middle Ages. Indeed, skeps were not popular in southern Italy, and we see evidence of this with the Exultet Rolls, a series illuminated manuscripts produced between 1000 and 1100. The Exultet Rolls included the text of a Catholic service that was held to praise the beeswax paschal candle, and many of the passages sing praises to the bees that produced the wax for the candle. It also includes illustrations of beehives used in Italy at the start of the High Medieval Age, the first illustrations of hives since ancient Egypt. Twenty of these manuscripts illustrate beehives. In general, these hives show that little had changed during the time since ancient Rome. Many of the hives were cylindrical or rectangular that were placed horizontally on a table or the ground. Some of the hives were stacked two to seven hives tall. Measurements based on the illustrations of the hives suggest that most were generally 2 to 3.5 times longer than wide or in diameter. The cylindrical hives were made from hollow logs, and possibly bark, whereas the rectangular hives were formed from small fennel sticks, or boards. One illustration shows two singular cylindrical hives on a platform with stout legs (Crane and Graham 1985; Kritsky 2017).

FIGURE 4.2 A bee niche or bee bole at Pluscarden Abbey near Elgin, Scotland. This bee niche was constructed around 1250 and is still used today. Photograph by Gene Kritsky.

Upright log or bark hives appear in manuscripts during the Late Medieval Age. It is difficult to determine the materials used, but some of the hives were placed on small tables and have a cut circular lid that is reminiscent of cork hives (Figure 4.3). Cork hives have the advantage of being very lightweight and made from renewable materials. To fashion a cork hive, the beekeeper would cut the cork bark away from the tree and roll it into a tube. The seam could be held together with stitches made of plant materials or held together with pegs. The lid was also made from cork held together in the same manner as the seam (Kritsky 2016).

Forest, or tree, beekeeping thrived in eastern Europe during the Middle Ages. Beekeepers would select a tall straight tree, preferably with a small natural cavity made by woodpeckers, which the beekeepers would enlarge by cutting

FIGURE 4.3 A cork hive showing the how the seam was stitched together. Photograph by Gene Kritsky.

a long rectangular opening in the bark, and carving out the interior. The piece of cut bark would be used to place over the cavity to provide protection and to conceal the bees from human thieves. The bees could be attracted to the cavity by honey that the beekeeper would have placed inside. If the tree was tall and straight enough, the beekeeper might create two or even three cavities in the tree to house up to three colonies, with the lowest hive being no less than 15 feet above the ground and the highest up to 75 feet up the tree. The honeycomb would be collected and placed in cloth bags for transport. If the tree should perish, or was in danger of falling, the hives could be saved by cutting the tree above and below each hive, and placing it in another tree or on the ground.

Tree beekeeping required the beekeeper to use ladders, ropes, chisels, saws, and drills as part of their apicultural equipment. Some later illustrations have the beekeeper using a crossbow to deal with the destruction of the bees by bears (Kritsky 2017).

Beekeeping practices can also be gleaned from illuminated manuscripts. One thirteenth-century illustration shows a beekeeper releasing a captured swarm from a cloth bag. The thirteenth-century Bodleian Bestiary mentions that bees will flee smoke (Kritsky 2017).

The primary purpose of beekeeping is to obtain honey, the only natural sweetener in the western world during the Medieval Age. The demand for honey made it a valuable commodity for trade, which is well documented during the Early Medieval Age. Charlemagne, also known as Charles the Great, unified much of western Europe and ruled as Emperor from 768 until 814. He issued a decree that two-thirds of the honey and wax produced on estates was to be paid to the Crown as a tax, which required every estate to have a beekeeper to oversee the production of honey and wax. As in the ancient world, honey in the age of Charlemagne was an important medicinal. A liver treatment involved the patient consuming a "spoonful of honey, vinegar, mustard, and 10 grains pepper" (Riché 1978).

King Alfred of Wessex, who reigned from 871 to 899, issued several laws dealing with bees and honey. One required that beekeepers had to announce when a swarm had emerged from a hive by ringing bells or making other loud noises to alert people to follow the swarm so it could be quickly captured. Moreover, any persons caught stealing bees would have to pay severe fines. The value of honey and wax during this period is underscored by the use of honey to pay rents. In 688, honey was measured by volume using a dolia, a large pottery vessel (Ransome 2004).

Bee products played an international role during the Early Medieval Age. In 879, King Alfred defeated the Viking army at the battle of Edington. After the war, trade was established between the Vikings and the English with the Vikings providing fish and furs in exchange for wool, wheat, malt, and honey. In eastern Europe, in what is present-day Ukraine and Belarus, honey, and particularly beeswax, made beekeeping a profitable occupation. Indeed, the demand for beeswax was such that it became a critically important raw material throughout Europe (Crane and Walker 1999; Kritsky 2017).

An unexpected use of beehives was their function as weapons that were especially effective against sieges. In 908, the Danes and Norwegians laid siege to the town of Chester. The inhabitants of Chester responded by first throwing large rocks on the invaders, who simply built fortifications to protect them from the rocks. The townspeople next poured boiling mead, which was made from honey and water, on the Danes who responded by covering their fortifications with hides. Finally, the inhabitants gathered the beehives in the town, and threw them down on the Danes. According to an old Irish account, "The Danes were prevented from moving their hands or the legs owing to the number of the bees that stung them" and the siege was broken. Bees factored in a similar siege in Germany thirty years later, when many beehives were flung out to attack horsemen. The bees were so angered by the disturbance that they "rushed out stinging the horses in such madness that their riders could not control them" (Herrod-Hempsall 1937).

The tactical use of beehives as weapons of mass destruction was also employed by the Muslims during the First Crusade in the first century of the High Medieval Age. The action occurred at Marrah, fifty miles from Antioch. This time the Muslims used "military engines" to sling the hives at their attackers. This was a considerable escalation in the use of beehives by using a catapult to send the hives some distance from the walls that were protecting the Muslims. By the Third Crusade, King Richard I of England carried beehives with him on the ships used to cross the Mediterranean to reach the Holy Land. This time the hives were used against Saracens living within the protective walls of Jerusalem (Anonymous 1902).

The extent of beekeeping in England at the start of the High Medieval Ages is well documented in the Domesday Book, a survey conducted after 1066 when William the Conqueror defeated King Harold at the Battle of Hastings. The Domesday Book, which was completed in 1088, was a survey of lands and property to determine what was owed to the Crown. It also documented the number of beekeepers and hives in each shire. The survey indicated that 30 percent of the landholdings in Essex County had bees, with each holding home to four to five hives. This totaled 601 hives in the county. It also appeared that the western counties used wicker skeps, whereas the eastern counties used larger straw skeps (Fraser 1958).

Other records provide details of the productivity of the hives. An account book dating to 1269–1270 for Beaulieu Abbey in Hampshire County, some ninety-five miles southwest of London, documents the economic impact of a large beekeeping operation. The account books shows that the Abbey often purchased empty straw skeps, which would be expected if the beekeepers were killing their bees to harvest the honey and if they had heavy winter losses. While honey was important for food, the Abbey required beeswax for liturgies, to produce the Easter candles that had to be made from pure beeswax, to seal casks, and to make votives. The votives were figurines of body parts, humans, and animals that were left at the tomb with the hope that it would effect a cure (Radford 1949). The Abbey's beekeeping operation produced 155 lbs (70 kg) of beeswax during 1269–1270. Honey yields were impressive when you consider that the records did not include the honey that was used by the monks. The year's production was 1,782 lbs (803 kg) of honey. Unfortunately, the records do not include the number of hives the Abbey had in production, but considering they were likely using straw skeps their honey production would suggest they had around three hundred hives (Vernon 1979).

Beeswax was also needed for sealing letters and other documents. Sealing wax was made from beeswax mixed with Venice turpentine, a resin from the European larch. It was usually colored with vermilion to give it a dark red color. The sealing wax would be applied to a folded letter to ensure the confidentiality of the contents. The wax was heated and allowed to drip where the letter would be opened, and then before it cooled was stamped with a signet ring that would identify the sender. In our age of cyber security, it is easy to think that these communications were easily opened. However, there were security systems that were used to determine if the seal had been broken. Most secure were seals that were stamped on both sides, which rendered them more difficult to remove without damaging the seal. Such documents were usually carried in a cloth bag to further protect the seal and assure the recipient that the letter had not been opened (Watson et al. 2015).

Abbeys also required quantities of beeswax for lost wax casting of chalices, reliquaries, crucifixes, candlesticks, and other religious artifacts. Lost wax casting was first used in antiquity as early as 3500 BCE and involves carving a beeswax model of the object to be cast into metal such as gold, silver, or bronze. This wax model is then used to make a mold, into which the molten metal would be poured. How this was accomplished in the High Medieval Age was described in the twelfth century by the Benedictine monk and metalsmith Roger of Helmarshausen Abbey in Germany. His metalworking knowledge was quite sophisticated as documented in his description for making the metal handles for a chalice (Hawthorne and Smith 1963).

> If you want to fix handles on the chalice, as soon as you have beaten and scraped it and before doing any other work on it, take some beeswax and form handles from it and carve on them snakes, animals, birds or foliage in any way you wish. On the top of each handle put a small piece of beeswax, rounded like a slender [candle] as long as your little finger but rather thicker at the top. This is called the sprue and you should join it with a hot iron. Then take some well-kneaded clay and carefully cover each handle separately so that all the hollows of the modeling are filled. When the handles are dry, carefully cover them again all over except for the top of the gate (the channel formed by the sprue); and do the same a third time.
>
> Afterwards put these molds near the coals and, when they are heated, pour out the beeswax. After it has been poured out, put the mold immediately into the fire, turning downwards the holes though which the beeswax came out, and leave them until they are red-hot like the coals. Immediately melt the silver and add to it a little Spanish brass … Take the molds away from

> the fire, stand them up firmly and pour in the silver in the same place from which you poured out the [melted beeswax].
>
> When they are cold, take away the clay and ... fix the handles to the chalice in their proper place.
>
> (Hawthorne and Smith 1963: 130)

Helmarshausen Abbey, where Brother Roger had his workshop, was also a well-known scriptorium, and Roger cast the metal components of illuminated manuscripts and reliquaries, some of which still survive.

Seals were also made through lost wax casting; however, the process to construct the seal model used to make the mold was different than that described by Brother Roger. During the Early Medieval Age, detailed carvings of the seal's design were carved in bone. Beeswax would then be pressed into the carving to make the wax model, which would then be cast into metal. In the National Museum of Ireland collection are bones with carved seal designs that have been found in abbeys and in wealthy residences that date between 700 and 1000 (Hunt 1980).

Many of the detailed large bronze or gold brooches that were produced in the Medieval Age were also made through lost wax casting. After the wax model had been carved, it would be pressed into soft clay. Another piece of clay would then be placed over the wax carving forming the back of a two-sided mold. The beeswax model, now sandwiched between the two pieces of clay, would be melted away, before the molted bronze or gold would be poured into the mold to produce the brooch (Hunt 1980).

Lost wax casting was also used for large-scale manufacture in the Medieval Age. Among the more impressive constructions were the large bronze doors that were made in 1015 for the Hildesheim Cathedral in Germany (Calkins 1979). The doors measured just over 15 feet tall and approximately 4 feet wide and 1.5 inches thick. Each door included eight scenes that were carved from large beeswax tablets which when completed were connected to each other using an iron frame for support, forming a massive single wax model for each door. The wax model was then covered with clay to make the mold, and molten bronze was poured into the mold forming the complete door. This technology must have seemed almost magical to many during the Medieval Age as suggested by an inscription on a lion's head door knob at Trier Cathedral that reads, "What wax created, fire has taken and bronze reproduced" (Toman 2007).

Highly ornate ecclesiastical objects, such as the Gloucester candlestick assembled between 1104 and 1113 and on display in the Victoria and Albert Museum in London, were cast in pieces. It is just shy of two feet tall and weighs over 10 lbs (4.5 kg). The sections include a three-footed triangular base, a main stem, and the pan that holds the candle and collects the melted wax. The main part of the candlestick is intricately carved with humans, animals, and plants. The three sections were held together with a copper tube. It was initially carved in beeswax and then cast in a copper alloy before being gilded (Hunt 1980) (Figure 4.4).

The industrial applications of lost wax casting during the Medieval Age were more complicated than simply carving a bas-relief of a biblical scene as evidenced by the manufacture of small bells. These bells, or cymbala, required a series of eight or nine bells that were the same size but were made to make each produce a different note. The trick to their manufacture was the weight of the beeswax. Brother Roger included instructions in his book. He wrote, "Whoever wants to make small bells sound in tune for playing should divide the wax for each one by weight. He should start to form the higher tones and descend until he reaches the lower ones" (Hawthorne and Smith 1963: 176). The process sounded complicated, but was relatively simple and when

FIGURE 4.4 The base of the Gloucester candlestick was made by the lost wax casting method. Photograph by Gene Kritsky.

completed would result in the bells covering a full octave. The trick involved carefully weighing out two equal pieces of beeswax, and then dividing one of the two pieces into eight parts. By adding different pieces of wax together, the metalsmith could obtain the precise amount of wax to produce bells that could ring the full octave. With the wax weighed out, they could be molded into the shape of the bell. Tin and copper would be melted down for casting. The subsequent small bell could be tuned by simply filing the bottom of the rim to produce a higher note or filing the rim on the outside of the bell to lower the tone. Clearly metalsmiths were also expert beeswax workers becoming sculptors by carving figures, preparing thin beeswax into sheets that would be used to cast censers, or measuring out the necessary amounts of wax to make a series of bells. Given the ecclesiastical demand for objects of worship there is little doubt why so many abbeys kept large numbers of beehives to provide the large quantity of beeswax that were necessary to cast these metal objects (Hawthorne and Smith 1963).

Beeswax had other metalworking applications by serving as a resist for etching patterns on swords or armor. Engraving involves cutting a design in metal and is a laborious process. An easier process used in the Late Medieval Age involved painting the helmet with a thin layer of beeswax. Once dry, the armorer could simply draw the design in the wax using a sharp tool. Drawing the design removed the wax from the surface of the helmet, breastplate, or other part of the armor exposing the iron. When the design was completed, the object would be washed with an acid solution that would etch the exposed surface of the armor. When completed, the beeswax would be removed and the surface rubbed with soot or lampblack, which would bring out more details of the design (Crane 1999; Houston 2016).

Other uses for beeswax during the Medieval Age include encaustic painting. Encaustic painting is thought to have originated in ancient Greece; however, the oldest known examples were made in Egypt during their Roman Era between 100 and 300. Encaustic painting involved mixing heated beeswax with various pigments. These "paints" were applied to a prepared board to create the image (Kritsky 2015). Encaustic painting continued into the Medieval Age as evidenced by icons that were created around 550 at St. Mary's Monastery in the Sinai. The method was later replaced by tempera painting during the eighth and ninth centuries. Tempera uses an emulsion of water, eggs (sometimes just the yolk), and occasionally honey. Unlike encaustic painting that creates rich colors, tempera painting involves several thin layers of paint that do not preserve rich colors, but are less prone to deterioration over time. Although encaustic fell out of favor in western Europe, it continued to be employed in the eastern Orthodox countries from 950 until 1500 (Encyclopedia of Fine Art 2018).

Beeswax was also used as an adhesive during the Medieval Age. The gemstones in Charlemagne's Iron Crown were affixed to the crown with a beeswax and clay mixture. This adhesive permitted modern researchers to accurately date the crown, which purportedly included a nail from the "True Cross." The beeswax in the adhesive was carbon dated to between 700 to 780 (Crocker 2018).

Another important application of beeswax that was a holdover from antiquity was the use of beeswax tablets to write out temporary documents (Figure 4.5). At a time when paper and other writing surfaces were at a premium, beeswax tablets were the method of choice to write out drafts. These tablets were made of wood with a recessed center, which would be filled with beeswax. Two tablets could be stitched together with leather strips that could be closed like a book, which would protect the wax and the information written on it from being damaged when the tablets were being carried. The use of such tablets during the Medieval Age is documented by the writings of the Abbot of the Abbey of St. Martin of Tournai, Hériman of Tournai (1095–1147), who chronicled the history of the Abbey initially writing on wax tablets (Nelson 1996).

For another example of how insects and their products could impact the medieval world, we only need to examine the workings of a scriptorium, a place where medieval manuscripts were produced. Scriptoriums were housed in abbeys and were often associated with metal workshops such as Friar

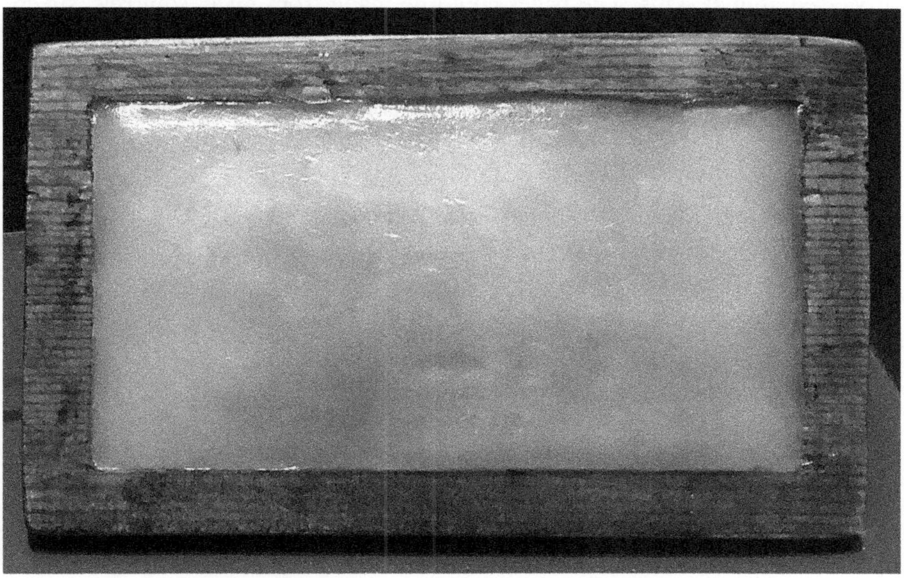

FIGURE 4.5 A beeswax writing tablet. Photograph by Gene Kritsky.

Roger's workshop at Helmarshausen Abbey. The process of making a medieval manuscript was an elaborate process as was described by one of the Exeter Book of Riddles.

> An enemy ended my life, took away
> My bodily strength; then he dipped me
> In water and drew me out again,
> And put me in the sun where I soon shed
> All my hair. The knife's sharp edge
> Bit into me once my blemishes had been scraped away;
> Fingers folded me and the bird's feather
> Often moved across my brown surface,
> Sprinkling useful drops; it swallowed the wood dye
> (Part of the stream) and again travelled over me,
> Leaving black tracks. Then a man bound me,
> He stretched skin over me and adorned me
> With gold; thus I am enriched by the wondrous work
> Of smiths, wound about with shining metal ....
>
> (Dormer 2012)

The pages of medieval books before 1400 were not made of paper as it had not yet been introduced into Europe. Instead, the pages were made of parchment, processed animal skins. The pages could be made from mice, squirrels, sheep, goats, and cattle. If the animal was a calf, the parchment was called "vellum."

The process to make parchment involved taking the skin from the animal and placing it in a flowing stream or river for a couple of days. It was necessary to place smooth rocks on the skin to keep it in place as it was washed. Next the skin was submerged in a barrel of lime and water for a week. This step was needed to help scrape the hair from the skin using wooden paddles. The hairless skin would then be scraped to remove any flesh that was still attached. After the hair had been removed, the skin would be put in fresh water for two days to remove the lime. The skin was then tied into a frame for stretching and covered with a chalk dust paste that permitted it to be stretched even tighter. During this time, it would be shaved with a semi-circular knife, called a lunellum, until it reached the desired thickness. The skin would be removed from the frame once it was dry.

The dried skin would be inspected for blemishes that would have resulted from wounds the animal received during its life. These blemishes included insect bites that might lessen the parchment quality. If such blemishes were present the parchment would more likely be used for more utilitarian books rather than a religious text.

Preparing the parchment to receive ink necessitated that it be prepared by rubbing the surface with a pumice stone, brushing away the powder with a feather, and covering it with a sticky powder. The book's topic would impact the size of the pages, which in turn determined how the skin would be folded and cut. If the book was to be long, the skin would be folded several times and cut along the folded edges, producing several pages of the needed size.

With the parchment prepared the pages were sent to the scribes, who would transcribe the text on the page. Manuscripts with uniform lines of text were highly prized, which was accomplished by the scribe ruling the paper with a straight edge (see Figure 7.3) (Dormer 2012; Houston 2016).

Scribes used quills for writing the text. These quills were made from feathers which they prepared by soaking them in water and then hardening them by placing them in hot sand. The vane and downy barbs of the feather would be cut away and the tip would be trimmed to the desired nib size before using. The scribe would also prepare the ink, and a popular black ink was made from oak galls, also called oak nut galls or oak apples.

FIGURE 4.6 Oak nut galls used to produce gall nut black ink. Photograph by Gene Kritsky.

Oak galls are produced when the cynipid wasp (*Biorhiza pallida*) lays an egg in the twig of an oak (Figure 4.6). The egg hatches into a larva whose feeding stimulates the plant to produce a round growth or gall that encompasses the larva. These galls are useful in making ink because they contain tannins, complex phenols that have been used to tan hides, make dyes, and inks.

Gall nut ink required only four ingredients: gall nuts for the tannins; gallic acid, water, or wine to act as a solvent; vitriol to be a source of iron; and gum Arabic as a binder. Gall nuts were soaked in water which produced a clear solution, but when mixed with iron salts such as vitriol turned a purplish-black color which got even darker as it aged (Thompson 1956). The resulting ink was amazingly permanent, enduring over fifteen hundred years. Many important documents were written in gall nut ink, including the Magna Carta which was signed in 1215.

Another black ink used was lampblack, which was made from soot made from a suspension of soot, honey, gum Arabic, and water. Thus, insect products were involved with both types of black ink (Meri 2006).

Scribes would write the text prior to doing any illustrations. This required the scribe to have an understanding where an illustration would be placed on the page and to make certain that there would be space for the illustrator to conduct his or her work. The illustrator would sometimes make preliminary sketches on wax tablets, which would then be copied on the page (Dormer 2012). There were several pigments that could be used, but kermes red was common. Kermes red was made from dried female scale insects, mostly *Kermes vermilio,* which feed on evergreen oaks in the Mediterranean region of southern Europe. First used in antiquity, this source of red pigment was used for dying fabrics in addition to being used for the red color in medieval manuscripts (Donkin 1977). The black pigments used to create illuminations were the same inks used in the texts (lampblack and gall nut ink).

With the pages completed, the book was bound with wooden boards that were covered in leather and held tight with a clasp or tied. The leather could be tooled using heated tools that were cooled with beeswax (Szimai 1999). If the book was to serve an ecclesiastical function, it would be sent to the metalworking workshop for ornate book covers which could include elements made by lost wax casting. Any precious stones would be set using beeswax as an adhesive (Dormer 2012).

All the processes described in the riddle about books involved insects in some manner. The examination of the parchment for insect blemishes might determine if the parchment would be for ecclesiastical purpose or not. The inks used required insects to make the galls that were used for the text. The reds and blacks used in illuminations required insect products. Finally, stones set in the ornate covers required beeswax to secure them in their settings. The very books that survive from the Medieval Age that document beekeeping techniques and our understanding of insect diversity required insect products for their production.

SILK

Silk is an insect product that changed the course of history and the economies of nations during the Medieval Age. Silk is produced by caterpillars of the species *Bombyx mori*. Legend has it that silk was discovered four millennia ago by the Chinese princess, Si-ling, who found a silkworm cocoon in her garden. While she was examining the cocoon she accidentally dropped it into her hot tea. After getting it out of her tea she noticed it has softened and was made from a single very long thread. Allegedly, she attempted to weave some of the thread and was so impressed with the results that she had her servants rear these caterpillars to obtain more cocoons (Clausen 1954; Cherry 1987).

Whether the Si-ling story is more folklore than truth is not clear. What is known is that silk production was discovered in China around 2700 BCE during the Huang-Di Dynasty (Huang-Di was married to Si-ling), and the Chinese endeavored to keep the process of silk production secret, especially from the nations to the west for over three thousand years (Cherry 1987).

Silk is produced by several insects and spiders, but it is the *Bombyx mori* caterpillar that produced silk that was successfully spun into thread for use in making the silk fabric. The silk is produced by labial glands that account for a quarter of the weight of the caterpillar (Grimaldi and Engel 2005). Silk is composed of two proteins, sericin and fibroin, of which fibroin forms silk's core around which sericin is wrapped. As the glands produce the liquid silk, it is issued out of the glands via spinnerets. The liquid silk will turn into solid silk when it is contact with the air (Clausen 1954; Millward 2013).

The adult female silkworm moth produces between 200 and 350 eggs, which will hatch into caterpillars that feed on mulberry leaves for thirty-five days. Then the caterpillars start producing silk that will enclose their cocoons, and when completed the cocoon will be enclosed with about 4,000 feet of silk! The completed cocoon will be immersed in hot water to kill the developing pupa inside and to soften the silk. The softened cocoon is carefully brushed to find the end of the silken fiber, which will then be reeled onto a wheel. This fiber is too delicate for weaving, so it will be twisted with usually five to seven fibers to produce silk thread, meaning that five to seven caterpillars were needed to make a single thread, and the silk from two thousand cocoons would be needed to produce a single garment (McGavin 2001).

In 120 BCE, an alliance between the Parthian Empire and China imported the first Chinese silk to Europe. The silk fabrics coming into Parthia were not the delicate silky fabric we think of today, but were much more coarsely woven. Recognizing an opportunity, the Parthians first sent this silk to Eyre, Sidon, and other locales where the fabric was unraveled and rewoven into the thin cloth that was so prized. The process was so effective that the Parthians exported this rewoven silk back to China (Thorley 1971).

The importation of silk to the west helped to expand the Silk Road, a trade route that connected China with the Mediterranean and Rome. When Rome's Emperor Augustus died in 14 CE, the demand for silk from China was high, and the Chinese were able to maintain this monopoly on silk production for another three hundred years.

The secret of sericulture made its way to Japan via Korea by the early third century (Cherry 1987). It spread westward in the third century, allegedly smuggled into India by a Chinese princess who hid silkworm eggs and mulberry seeds in her headdress (Cherry 1987; Millward 2013). It was only a matter of time before Byzantium would gain this technology.

When Justinian became Emperor in 527 most of the silk was imported from Persia which started producing their own silk in the fifth century. This silk was expensive as it was dyed and designs were woven into the fabric, which added to the costs. Moreover, relations with Persia were tense and exacerbated by the silk trade. Clearly, there would be economic and political benefits if silk could be produced within Byzantium. It was Byzantine historian Procopius who recorded what happened in 552:

> About the same time there came from India certain monks; and when they had satisfied Justinian Augustus that the Romans no longer should buy silk from the Persians, they promised the emperor in an interview that they would provide the materials for making silk so that never should the Romans seek business of this kind from their enemy the Persians, or from any other people whatsoever. They said that they were formerly in Serinda, which they call the region frequented by the people of the Indies, and there they learned perfectly the art of making silk. Moreover, to the emperor who plied them with many questions as to whether he might have the secret, the monks replied that certain worms were manufacturers of silk, nature itself forcing them to keep always at work; the worms could certainly not be brought here alive, but they could be grown easily and without difficulty; the eggs of single hatchings are innumerable; as soon as they are laid men cover them with dung and keep them warm for as long as it is necessary so that they produce insects. When they had announced these tidings, led on by liberal promises of the emperor to prove the fact, they returned to India. When they had brought the eggs to Byzantium, the method having been learned, as I have said, they changed them by metamorphosis into worms which feed on the leaves of mulberry. Thus began the art of making silk from that time on in the Roman Empire.
>
> (Halsall 1998)

The monks returned to China where they obtained silkworm eggs. They concealed the eggs in bamboo poles to carry back to the west (Millward 2013). They may also have returned with small mulberry shrubs and root balls.

A slightly different account was recorded by the historian Theophanes who wrote, "During the reign of Justinian a person showed at Byzantium the birth of silkworms. This had been unknown to the Romans up to that time. Before he left the country of the Seres [China] he took the eggs of the silkworms in a hollow stick and brought them safely to Byzantium" (Forbes 1956: 58). The acquisition of silkworms and knowledge of their husbandry meant that the Chinese monopoly would be broken. It also meant that the intermediaries along the Silk Road would not be in a position to add to the cost of silk by extracting duty fees to silk passing through their territories.

Silk was a major trade commodity and demand for it was high. It was used as a currency and for diplomatic gifts. It was second only to gold in its value and there were times when this insect product was in greater demand than gold. By the reign of Justinian's successor, Justin II, Berytus, now known as Beirut, had become a major producer of silk. Aiding the process were sumptuary laws that Justinian issued dictating who was entitled to wear purple dyed silk. The purple dye was obtained from murex, a marine snail collected from the eastern Mediterranean Sea. Prior to these laws, only the wealthy could afford to purchase and wear purple silk. However, the Justinian sumptuary law made purple silk with gold embroidery a symbol of royalty and the religious elite (Hunt 2007). The Code of Justinian also controlled the production of silk and the purple dye that made such clothing a diplomatic tool (Liu 2001). How the Emperor accomplished his goal of taking over the silk trade was described by Procopius in the mid-sixth century.

> By licensing monopolies of nearly all kinds of wares, these rulers daily oppressed the purchasers; the sale of [silk] clothes was the only thing they left untouched, and even in this case they contrived the following scheme. Cloaks of silk had long been made in [Beirut] and Tyre, in Phoenicia. Merchants who dwelt in these, and all the artisans and workers connected with the trade, had settled there in early times, and from these cities this trade had spread throughout the earth. But during the reign of Justinian, those in this business at Constantinople and in the other cities, raised the price of [silk] garments: claiming that the price for [silk] had been raised by the Persians, and that the import duties to Roman territory were also higher.
>
> The emperor pretending to be incensed at this, proclaimed by edict that [silk] could not be sold for more than eight gold coins a pound; and the punishment for disobeying this law was the confiscation of the transgressor's property. This seemed to everybody impossible and futile. For it was not practicable for the merchants who imported silk at a higher price, to sell it to their customers for less. Consequently, they decided to stop dealing in it at all, and privately got rid of their present stock as best they could, selling it to such notables as took pleasure in throwing away their money for such finery, or thought they had to wear it.

> The Empress ... immediately confiscated these persons' wares, fining them a centenary in addition. Now the imperial treasurer is to be in charge of all matters connected with [silk] trade, [who] ruled that all should obey the letter of the law, while he ordered the silk makers to work for him. And this was no secret, for he sold colored silk in the Forum at six gold pieces an ounce, while for the imperial [purple] dyed [silk] ... he charged more than twenty-four.
>
> In this way he got much money for the emperor and more, quietly, for himself; and the custom he started continues to this day, the treasurer being admittedly the sole silk merchant and controller of this trade.
>
> The former dealers in silk in Constantinople and every other city, by sea and by land, were naturally heavily damaged. ... Artisans and mechanics consequently left the country and fled to Persia. Only the imperial treasurer could transact this business, giving a share of the profits aforesaid to the emperor, and himself taking most of them, fattening on the public calamity.
>
> (Procopius 2007: 121–3)

Silk production was controlled through the establishment of silk guilds, which included the craftsmen who produced the fabrics, and guilds of men involved with the sale of silk. The Medieval Book of Prefect or Eparch, which dates to the ninth and tenth centuries, includes several regulations for silk merchants, silk importers, raw silk merchants, and silk dyers. The sophistication of the guild's regulations is seen in some of the regulations. For example, section five reads, "To obtain admittance to the guild of silk merchants five members of the craft must testify to the eparch that the candidate is a person worthy to exercise the craft. He shall then be admitted to it, he shall open a shop, and carry on business. His entrance fee to the guild is six nomismata." The first regulation for a silk dyer reads, "Dyers are forbidden to make up the purple of the so-called prohibited grades, that is to say in the series of great mantles, including those of self color or those where the purple alternates with dark green or yellow in half-tint" (Atchison 2023).

Silk took on many uses during the Early Medieval Period. Diplomats to Constantinople were given gifts of silk. Silk was used for ecclesiastical vestments, it was used to wrap relics and crucifixes, and to cover books. Silk also played a role to rally troops going into battle. Silk standards that announced the allegiance of the troops were easily transported and displayed during battles (Muthesius 1992). Silk shrouds were also used in burials.

The desire for silk during the Medieval Age encouraged vast trading networks as revealed by ship burials, a Viking practice that placed the remains of the deceased and grave goods in a buried ship. One such burial excavated in 1904 was used in the internment of two women. Among the grave goods found were silk garments. Other finds included strips of silk that likely were used to decorate more traditional clothing. Dendrochronology of beams used in the ship placed the burial around the year 834. The Vikings also used silk as a symbol of rank, as suggested by the grave also including jewelry and other objects of value. Silk has been found in burials of men and women, although the number of silken objects is greater in women's burials (Vedeler 2014).

The Byzantine monopoly of silk production was broken after the Muslims defeated Persia and came into the knowledge of silk production. They, in turn, spread sericulture to Sicily and Spain. Eventually, silk production spread further into Italy; and by the thirteenth century, silk from Europe was competing with Byzantium silk, changing the economies of several regions in Europe (Snell 2017).

CHAPTER FIVE

Insects in Mythology and Religion

VÍCTOR J. MONSERRAT AND VAZRICK NAZARI

The origins of universal beliefs, myths, and ancient symbols in the West date back to the people of Old Europe (Ransome 1937; Hall 1974; Hogue 1975, 1987; Gimbutas 1991, 1996; Moser 1998; Bruce-Mitford 2000; Kritsky and Cherry 2000; Impelluso 2003; Monserrat and Aguilar 2007; Monserrat 2010, 2011a; Monserrat and Melic 2012, etc.). In the journey towards western civilization, and through medieval Christianity, the mythology and iconography related to the animal world were later influenced by two main sources: the *Classical* and the *Sacred* texts.

Myths and legends from the Classical texts dating back to Mesopotamian, Assyrian-Persian, and Hindu cultures first gave birth to the curious, intellectual, and restless Greeks, and subsequently to the eclectic, imaginative, and tenacious Romans. Rome took the Greek heritage and added many new elements to it, either from their own culture, or from those acquired via fair trade. Soldiers returning from conquered lands, especially from the Mediterranean (e.g., Phoenician-Punic world and particularly ancient Egypt) introduced many new cultural elements into Roman Society (Cumont 1929; Davies and Kathirithamby 1954; Lauwerier 1988; Dalley 1991, 1998; Caquot et al. 1993; Charbonneau-Lassay 1996; Mariño Ferro 1996; Eliade 1999; Bry 2000; Roaf 2000; Clauss and Gordon 2001; Blázquez et al. 2011; Monserrat 2011b, 2012a, 2012b, 2013a, 2013b, 2013c).

On the other hand, after the Edict of Milan (313 CE) and the establishment of Christianity as the official state religion by Theodosius (391 CE), and after the fall of the Roman Empire (476 CE) and the subsequent reorganization of Christianity in Europe and its progressive expansion, the Sacred texts of Christianity became the main reference and source of its rules, laws, and beliefs. These included, among others, the Revealed texts (Bible) with their Mesopotamian influence, the Patristic texts, as well as the texts of the Fathers of the Latin Church (St. Augustine, St. Gregory, St. Ambrose, and St. Jerome), and the texts of the Fathers of the Greek Church (St. Athanasius, St. Basil, St. Gregory, and St. John Chrysostom), which in both cases had significant Greco-Roman influence.

This chapter focuses on the zoological and entomological elements assumed by the West and by medieval Christendom represented in their cultural ideology, mythology, artistic, and literary manifestations, and religion. Throughout the Medieval Period, the ancient Greco-Roman influence gradually faded away, accommodating for the Sacred texts, especially the Old Testament and the New Testament (Birdsong 1934; Bruce 1958; Heider 1995; Kritsky 1997; Torralba Burrial 1998; Page 2007).

Irrespective of the source, the vast majority of the cultural and religious elements related to insects were inherited by Christianity in the West during the Middle Ages (Charbonneau-Lassay 1996). Over time this legacy evolved and incorporated new elements, parts of which still remain in numerous beliefs and popular manifestations until today. To understand the origins and the mentality behind the presence of these elements in the ideology of medieval European Christendom, we will focus on information obtained from these sources (Classical and Sacred texts, especially the Old Testament and the New Testament).

Aside from ancient myths and beliefs among European peoples, and the sparse and cryptic early Christian iconography in Roman times (Gough 1973; Grabar 1980; Beckwith 1985, 1997; Grant 1999), most of the surviving information about the animals and monstrous beings in the European Middle Ages comes mainly from the Roman World (Bayet 1971; Réau 1997a, 1997b, 1998, 1999, 2000; Carmona Muela 1998, 2000; Mittman 2006; Monserrat 2011b, 2012a, 2013c, etc.). It is true that Roman mythology is partially rooted in Etruscan beliefs; however, they also inherited from the Greeks, who in turn had adopted from the Eastern legends, mostly from Assyria, Persia, and India (Buren 1937–1939; Wittkower 1979; Beavis 1988; Graves 1996; Kritsky and Cherry 2000; Blázquez et al. 2011; Monserrat 2012b), particularly the *Mahabharata*, to which Herodotus first gave several references (Monserrat 2011b).

Stories of authors like Deimachus, Seleucid ambassador in the Indian court, or that of Ctesias of Cnidus, a historian and physician in the court of Artaxerxes II Mnemon of Persia in early fourth century BCE, were influential in shaping the European medieval imagination. Ctesias' story *Indica* (mainly surviving through the abbreviated version of Photios, patriarch of Constantinople in the ninth century), refers to India as "The wonderland with fabulous beings": sciapods or men with one leg, baboons or men with dogs' heads, men without heads, men with eight fingers on their hands and feet, others with huge ears, giants, pygmies, manticores, or men with a lion's body and scorpion tail, unicorns, griffins, chimeras, locusts, harpies, etc. (see Figures 5.1–5.6). This list of men and fantastic creatures was expanded by Megasthenes, who was appointed by Seleucus Nicator (heir to the Asian empire of Alexander) as ambassador to the court in Pataliputra of Sandracottus until 303 BCE. The surviving fragments and citations of Megasthenes's work by authors such as Diodorus, Strabo, Pliny, and Arrian or Elio, also introduce fantastical elements, such as snakes with bat wings, huge winged scorpions, gold-digging ants, antipodes (men with heels forward and toes back), men without nostrils or a functional mouth, hyperboreans who lived a thousand years, and men with dogs' ears and one eye (Gerhardt 1965; Clair 1967; Barber and Riches 1971; Hays 1973; Baur 1974; Rowland 1974; Farson and Hall 1975; Izquierdo and Le Meaux 2002; Campagne and Campagne 2005; Werness 2006, etc.).

FIGURE 5.1 Mosaic pavement of Santa Maria e Donato (twelfth century) with decorative scorpion tails (griffins/basilisks), Murano, Venice (Italy). Photograph by Victor L. Monserrat.

FIGURE 5.2 Dragon wearing scorpion tail, capital from Church of San Pedro (twelfth century), Fraga, Huesca (Spain). Photograph by Victor L. Monserrat.

FIGURE 5.3 Romanesque capital with fantastic figures wearing scorpion tail, Church of the Madalena (twelfth century), Zamora (Spain). Photograph by Victor L. Monserrat.

FIGURE 5.4 Lion wearing scorpion tail, tympani of Church of Asunción de María (twelfth century), Bossòst, Lerida (Spain). Photograph by Victor L. Monserrat.

FIGURE 5.5 Harpies in capital in the cloister of the Monastery of Santo Domingo (twelfth century), Silos, Burgos (Spain). Photograph by Victor L. Monserrat.

FIGURE 5.6 Basilisk with scorpion tail, Parish Church of Aínsa, Huesca (Spain). Photograph by Victor L. Monserrat.

These beings appear in many Classical texts, such as works of Aristotle, *The Labors of Hercules*, *The Aeneid*, *The Iliad*, *The Odyssey*, *The Metamorphoses*, and *The Natural History* of Pliny (Steel et al. 1999; Beullens 2007; De Leemans and Klemm 2007). This wealth of exotic and fantastical elements also inspired the authors of the books of the Bible, and consequently shaped the iconography of Christianity and the religious and imaginative minds of Europeans in the Middle Ages (Figures 5.1–5.6). The information in these Classical texts has also contributed to animals, including insects, in western iconography, especially in the Middle Ages.

The main sources of entomological iconography in medieval Christianity are the Sacred texts of the Bible, mainly the Old Testament, and to a lesser extent the New Testament, where we find numerous references to arthropods: locusts (39:20), moths (4:19, 13:28, 27:18), ant-lions (4:11, 28), fly larvae (7:5, 17:69, 21:26, 24:20), bees and honey (20:17), and spiders (8:14–15, 18, 27) (Birdsong 1934; Bruce 1958; Rowland 1974; Friedmann 1980; Heider 1995; Kritsky 1997; Réau 1997a, 1997b, 1998, 1999, 2000; Page 2007).

Entomological references to arthropods are also abundant in other Revealed texts (Birdsong 1934; Bruce 1958; Heider 1995; Kritsky 1997). The Bible makes ninety-eight references to bees, beetles, fleas, flies, mosquitoes, scorpions, wasps, grasshoppers, lice, locusts, moths, and palm worms, and some whose identities are unclear, like the "gnat" quoted in the third plague of Egypt, "All the dust throughout the land of Egypt became gnats" (Exodus 7:16–19, Psalms 105:31), and Matthew 23:24 (perhaps psychodids or some other variety of mosquitoes). There are also animals—arthropod hybrids, such as the scorpion-tailed locusts (Revelations 12:1–12).

Three of the ten Plagues of Egypt (מכות מצרים or *Makot Mitzrayim*) directly involved insects: the third with *Kinim*, variously translated as mosquitoes, fleas, or lice; fourth with horsefly or gengens; and eighth with locusts. Two or three others were somewhat indirectly related to insects: the fifth and sixth by ulcers and pustules originating in people and livestock (Exodus 7:14–25, 8:16–32, 9:1–12, 10:1–20). Some of them, like "throughout Egypt the land was ruined by the flies" in the third plague, or when God instructed Moses, "Tell Aaron, 'Stretch out your staff and strike the dust of the ground,' and throughout the land of Egypt the dust will become gnats ..." have not been identified yet but they perhaps refer to sand-flies (Psychodidae). But the most referred-to of those insect are the pests, specifically locusts (Insecta, Orthoptera: Acridiidae, *Anacrydium aegypcium*, and *Schistocerca gregaria*). This makes sense, given the farming and livestock-keeping lives of the Semites or the Chosen People who were always worried that the locusts will diminish or destroy their crops, especially after exodus and upon settling in the Promised Land. The idea that locusts can be used at any time as a ruthless and cruel "divine punishment" is well-reflected in the Apocalypse of St. John (Monserrat 2014). Clouds of locusts or their effects are also mentioned in other texts:

> "Before them fire devours, behind them a flame blazes; Before them the land is like the garden of Eden, behind them, a desert waste" (Joel 2:3); "I will repay you for the years the locusts have eaten, the great locust and the young locust, the other locusts and the locust swarm, my great army that I sent among you" (Joel 2:25); "Swarms of locusts will take over all your trees and the crops of your land" (Deuteronomy 28:42); "When I shut up the heavens so that there is no rain, or command locusts to devour the land or send a plague among my people" (2 Chronicles 7:13); "What the locust swarm has left, the great locusts have eaten; what the great locusts have left, the young locusts have eaten; what the young locusts have left, other locusts have eaten" (Joel 1:4); and so on (e.g., Joel 1:2–20, 2:1–11, 15–20; 2 Chronicles 6:28; 1 Kings 8:37; Book of Wisdom 11:15, 12:24, 15:18; Isaiah 33:4; Amos 4:9, 7:1–3; Nahum 3:15–17b; Psalms 22:17).

In addition, there are citations concerning the ten plagues (the eighth of locusts) sent by Yahweh against Egypt for not allowing the departure of the

Israelites (Exodus 10:1–20; Psalms 78:46, 105:34; Book of Wisdom 16:9; Judges 6:5, Luke 10:19, etc.). Reference to evil locusts can also be found among the Mesopotamian Bestiaries, as well as in medieval ones (Robinson 1961; McCulloch 1962; Clair 1967; Barber and Riches 1971; Clébert 1971; Mode 1975; Henderson 1982; Diekstra 1985; Yapp 1985; Roaf 2000; Resl 2007a, 2007b). Locusts were "destined" to do evil against those who separate from the "right way," and in the Christendom those include bad priests, false prophets, devils, and false Christian opponents of the Church.

Perhaps insects started on the "wrong foot" in their career with the Chosen People and thus in the Christian West, although curiously, and as a counterpart to these apocalyptic events, these destructive insects were interpreted as a symbol of the converted pagans by some (Proverbs 30:27). Locusts were considered "pure" and edible animals: "You may eat all winged creatures that crawl on all fours and also have two to jump on the earth" (Leviticus 11:21); "these you may eat; the locust, the beetle, the bald locust and grasshopper" (Leviticus 11:22). Acridophagy was common among the people of the Middle East (Defoliart 1989). John the Baptist even followed a very entomological diet: apparently, he ate grasshoppers in the desert with wild honey (Matthew 3:4: "and his food was locust and wild honey"). He is therefore associated with locusts as well as honey combs, which alongside other food stuff are cited with some frequency (Leviticus 11:20–2; Leviticus 20:25; Deuteronomy 14:19). Locusts were even considered one of the "four wise animals" (Proverbs 30:27). However, these animals acquire their most demonic, terrifying, and destructive significance in the Apocalypse of St. John (9.3, 9.7–12), leaving a long-lasting negative impact on the reputation of nearly all insects, including harmless ones (Monserrat 2014). In the Apocalypse of St. John, the term "pest" is used fourteen times in different contexts, mainly as punishment, while it is mentioned fifteen times in Exodus.

Apart from locusts, other insects were sometimes considered "pests" by livestock owners. For example, many horseflies, flies, and mosquitoes (Diptera) are harmful to cattle, and on occasion, also troublesome for people. Dipterans are often associated with putrefaction of corpses and death, because of the diet of some fly larvae, often simply called "worms." Metaphors, curses, omens, and references linked to this fact are common: "Worms will feed on his decaying body" (Sirach 19:3); "Side by side they lie in the dust, and worms cover them both" (Job 21:26); "My body is clothed with worms and scabs, my skin is broken and festering" (Job 7:5). In relation to the third plague of Egypt, "… the gnats were on people and animals everywhere" (Exodus 8:16–19, Psalms 105:31); and the fourth plague, "throughout Egypt the land was ruined by the flies" (Exodus 8:20–32, Psalms 78:45, Book of Wisdom 16:9). Sometimes it is quite expressive: "They swarmed around me like bees, but they were consumed as quickly as burning thorns; in the name of the Lord I cut them down" (Psalms 118:12). Worms (larvae of necrophilous flies) are icons of death and putrefaction since they kill instead of giving life, and with the serpent ("the great worm"). Worms, much like flies, are antithetical exponents of salvation and eternal life, and are also associated with demons, temptation, and death.

There are several other biblical beings related to insects (Keel and Uehlinger 1998; Leick 1998; Wyatt 1999). One such case is the Beelzebub, Semitic deity of the cities of Beel, Ekron, or Baal, whose presence in writing dates back to Mesopotamia, and mentioned in the Old Testament as "Zebub" (Isaiah 7:18, Ecclesiastes 10:1). The name Beelzebub is derived from *Baal Zebub*, or more accurately Ba'al Z'vûv, in Hebrew זבוב בעל with many slight variations. Beelzebub was also referred to as the "Lord of the flies" in Mesopotamia, "the dreaded demon of death" among Persians, and that in turn was the name of a Philistine god Baal Sabaoth (deity of hosts), or "Prince of the flies," worshiped in biblical times in the cities of Ekron and Avaris. People worshiped these deities across the Mediterranean region, and their influence reached even Rome and the Roman armies.

In the Book of Isaiah, hell is described as being full of snakes, toads, spiders, and worms (e.g., 66:24). The impact of this citation is evident in the ideology and medieval iconography: scorpions and spiders are often referred to as demonic beings associated with temptation, pain, and purgatory (Figure 5.7) (Mode 1975; Mellinkoff 1985; Heider 1995; Carmona Muela 1998, etc.).

FIGURE 5.7 Taddeo di Bartolo, False Testimonies (1396), with scorpions on the body of the unfortunate condemned, Collegiata Santa Maria Assunta (1148), San Gimignano (Italy). Photograph by Victor L. Monserrat.

Aphids and scale insects (Homoptera) are also frequently cited (Psalms 105:34, 1 Kings 8:37, 2 Chronicles 6:28, Nahum 3:15–17b). They are cited five times in Exodus 16, when speaking of manna, the substance that saved the Chosen People from hunger during their forty years of wandering in the desert towards the Promised Land. Other potential references to manna (heavenly food, bread of angels, miraculous food, bread of life) also appear frequently in other texts (Psalms 78:24, Gospel of St. John 6:31–5, Revelation 2:17, etc.). Although there are different hypotheses on what manna really was (*Tamarix mannifera* secretions, the lichen *Lecanora esculenta*, fungi, etc.) (Bodenheimer 1947; Schimitschek 1980), it could also be argued that it was a rather sugary substance made of secretions of insects (cochineal *Trabutina mannipara*, Homoptera: Pseudococcidae).

Moths are also mentioned several times in the Bible, usually in a negative context as pests of stored goods or clothing (Psalms 39:11, Isaiah 51:8, Hosea 5:12, Matthew 6:19, John 5:2). Because of their life history (metamorphosis) and through the Greek mythology (*Psyche*), Lepidoptera were metaphorically associated with the immortality of the soul, the resurrection, and the good behavior to follow. Certain lepidopteran species, generically cited as moths (Job 4:19) and some beetles, called "sappers" or "woodworms" (Job 13:28) are among the negative elements in the entomology of medieval Christianity. Fruit flies and some harmful moths mentioned in the Old and the New Testament (Deuteronomy 28:39–40, Isaiah 50:9, John 5:2) are assignable to *Dacus oleae* (Diptera: Tephritidae), *Lobesia botrana* (Lepidoptera: Tortricidae), or *Tineola bissellielia* (Lepidoptera: Tineidae) (Birdsong 1934; Bruce 1958; Heider 1995; Kritsky 1997; Torralba Burrial 1998). In the New Testament there are proportionately fewer references to non-domestic animals, including some insects (Paul to Corinthians 11:2–3, Matthew 4:10, Zechariah 3:1–2, etc.). In general, insects are treated with contempt and permanently branded as "irrational" in the New Testament.

Unlike Mesopotamia, Mycenae, ancient Egypt, or the Greco-Roman world, where arthropods constituted a substantial part of the belief systems and mythologies and often acquired the nature of deities (e.g., Cicada of Minerva, bees of Zeus, grasshopper of Apollo, the Egyptian sacred scarab Khepri, etc.) (Ward 1902; Weiss 1927; Swift 1931; Tod 1939; Monserrat 2011b, 2012a, 2012b, 2013a, 2013b, 2013c), in the symbolism and ideology of European civilization since the spread of Christianity, insects and other arthropods have been relegated to the dark underworld of the demonic, sin, and evil, and this bad reputation has survived to modern days in the form of our phobias of insects (Tod 1939).

In this new morality endorsed by the ideology of the European continent, few arthropods were spared from the fire. We see overwhelmingly in the Sacred texts that arthropods associate with danger, demons, sins, and punishment. Therefore, insects were generally relegated to evil, vice, sin, death, darkness, and the demonic (Cloudsley-Thompson 1990, 2001; Sorabji 1993; Réau 1997a, 1997b, 1998, 1999, 2000; Monserrat and Melic 2012), or are lowered to insignificant and destructible animals, as in when David compares himself to a flea (Isaiah 24:15), or "those ... who are crushed more readily than a moth!" (Job 4:19). Insects are generally a reason for disgust and annoyance: "they worship the most loathsome animals" (Book of Wisdom 15:18); "the teeth of wild beasts, and scorpions and vipers, and the sword that punishes the ungodly with destruction" (Sirach 39:30); "My father chastised you with whips, but I will chastise you with scorpions" (1 Kings 12:11); "you sitting upon scorpions" (Ezekiel 2:6). In this demonization, of them all, the scorpion's reputation for deception and its association with betrayal was reason enough for it to be linked forever with the Jewish people (Bulard 1934, 1935).

On this basis, all sorts of arthropods appear in the Middle Ages, each more damaging and more evil, and they become increasingly associated with creatures from darkness. This is the very antithesis of the symbols that were used to dignify their characters and creeds, opening a huge door to the medieval inventiveness and imagination, where thousands of evil monsters with entomological characteristics appear (Figures 5.1–5.6). In all medieval iconography, the zenith of this phenomenon can perhaps be observed in the Lapidaries, as well as in the Bestiaries and *Physiologus* during the Romanesque and Gothic epochs. Popular ideology, with its firmly rooted superstitions and beliefs from periods prior to the Christianization, and other beliefs adopted from Greco-Roman world, only further enriched this entomological medieval ideology.

It is fair to note that there are also some kind references to insects. For example, in Proverbs of St. Augustine (6:6–8, 30:25–7), Psalms (118:12), the *Hexameron* of St. Basil, and John the Apostle (Baptist) who makes a surprising and charming reference to bed bugs: "I say unto you, O bugs, behave yourselves, one and all, and leave your abode for this night and remain quiet in one place, and keep your distance from the servants of God" (The tale of John and the bedbugs, Acts of John: 60). Butterflies and bees are very frequently represented next to flowers, especially in scenes dedicated to the Virgin Mary, reinforcing the idea of virginity, innocence, and purity (Verdon 2005). The presence of butterflies and bees associated with images of the Virgin Mary was so predominant that it extended to other people and non-religious figures, where the use of butterflies enhanced the purity and innocence of the character of the lady to whom the text referred. White flowers often decorated such scenes as a symbol of virtue, purity, and cleanliness. Ants and bees were also treated with more leniency due to their industrious nature.

Apart from the Revealed texts, there are also many references to insects among the Grandees of the Church. These should also be considered, as they contribute towards the medieval entomological amalgam.

Entomology as a science was rudimentary in medieval Europe (Morge 1973; Smith et al. 1973; Scarborough 1979). Plutarch and Pliny the Elder made extensive and unquestioned references to the works of Aristotle, which were the main source of knowledge throughout the Middle Ages (Kitchell and Resnick 1999; Beullens 2007; De Leemans and Klemm 2007; Resl 2007a, 2007b). But several medieval texts related to zoology were also produced, such as Etymologies (623) of Isidore of Seville (570–636) (Isidoro de Sevilla 2004). The theologian, philosopher and Dominican friar Albert von Bollstädt, known as Albertus Magnus (1193–1280) (Albert the Great 1987, 1999), wrote *De animalibus* (1255–1270), a book that was extensively cited by later authors who exerted enormous and everlasting influence throughout the Middle Ages, not only on entomology, but also on science, literature, and decorative arts (Hays 1973; Beullens 2007; De Leemans and Klemm 2007; Resl 2007a, 2007b).

A prime example of medieval works related to zoology and entomology that greatly influenced the entomological mentality and medieval iconography is *Etymologies* (*Etymologiae* or *Originum sive Etymologiarum libri viginti*) (627–630) by Isidore of Seville (Hispalenses Isidorus) (570–636). This Latin work includes a compendium of all branches of knowledge, and within each branch, topics in alphabetical order from the Rhetoric and Grammar (Books I, II), Mathematics, and the Laws (Book IV, V) to the stones and metals (Book XVI), animals (end of Book XI and Book XII), boats (Book XIX), domestic chores and tools (Book XX), etc. Agronomy is also treated (Book XVII), and Book XII is dedicated to animals under the Virgilian-Aristotelian-Plinian vision of the time. It seems that Isidore of Seville was more interested in etymology than entomology, designating or maintaining terms such as *Pulex* for flea because it kept the classical belief that they were born from dust (*pulvis*), or *Pediculus* for louse for its small legs, without providing little new entomological insights. The book (especially Book XII, chapter V "De vermibus" and chapter VII "De volatibus diminutis" and "De transformatis") makes many references to worms, fleas, lice, and other small volatile creatures, and overall, twenty-seven insects are cited. It declares the origin of bees from dead cattle, beetles from horses, grasshoppers from mules, and scorpions from crabs.

The book gives us some idea about Isidore of Seville's vision of arthropods. He quotes Ovid on Scorpions:

> [When crabs decay, they transform into scorpions]; If you take its curved arms from a crab on the shore a scorpion will emerge and threaten with its hooked tail ... the scorpion are vermin of the earth, armed with a sting; it

> stings with its tail and infuses venom through the bow-shaped wound. It is characteristic of the scorpion that it does not attack the palm of the hand.
>
> (Book XI, 4: 3, and XII, 5: 4, 6: 17)

On spiders, he says: "Spiders (aranea) are vermin of the air (aer), named from the air that is their nourishment. They spin out a long thread from their little body and, constantly attentive to their webs, never leave off working on them, maintaining a perpetual suspension in their own piece of craftsmanship" (Book XII, 5: 2). And says of ants:

> The ant (formica) is so named because it "carries bits of grain" (fert micas farris). It has great shrewdness, for it provides for the future and prepares during the summer what it consumes in the winter; during the harvest it selects the wheat and does not touch the barley. When it rains on the ant's grain, the ant throws it out. It is said that in Ethiopia there are ants in the shape of dogs, who dig up golden sand with their feet – they guard this sand lest anyone carry it off, and when they chase something, they pursue it to death.
>
> (Book XII, 3: 9)

His vision of crabs is not any better, since he adopts the views of Ovid and Pliny in thinking that crabs are predecessors of scorpions (see above) and "are animals from evil, carriers of cruelty, greed and deceit" (Werness 2006). He mentions the terendons (termites), the silkworms, caterpillars, the Cantharis beetle and its effect on human skin, and says centipedes are born under the stones of the earth. He has a more positive opinion about ants, as well as crickets with their peculiar method of hunting with an ant tied to a hair. In Chapter 8, "Small Flying Creatures," he talks about bees (born without legs), wasps, hornets (originating from the rotten horses), buprestids (who deceive oxen on pasture), ladybugs (shining when they walk and fly), cockroaches (who can't stand the light), butterflies (arising from their droppings), cicadas (born from sputum of cuckoos), flies that revive after drowning, mosquitoes (drinking blood) and other midges (*sciniphes*) which he attributes to the third plague of Egypt (Exodus 8:16), bibions (born from wine), horseflies (very irritating) (Exodus 8:29), and weevils (*gurgulio*).

Many Greco-Roman ideas and concepts related to insects, especially those from Aristotle, Pliny, and Aelian, survived through the Classical texts into the Medieval Period. The enigmatic ichneumon (Egyptian mongoose) which lived on the banks of the Nile and was with the enemy of the crocodile hydra, and thus ended up being a symbol of Christ, is an example (Book XII, 17). But we hear more about the bees. Bees reproduced parthenogenetically and thus from ancient times were associated with virginity, and therefore to the Virgin Mary. This was a decisive contribution of Isidore of Seville, who in *Etymologies* indicates:

> [Bees emerge out of the rotted flesh of calves] ... Bees (apis) are so named either because they cling to each other with their feet (pes), or because they are born without feet (cf. a-, "without"), for they develop feet and wings afterwards. These animals, skilful at the task of creating honey, live in allocated dwellings; they construct their homes with indescribable skill; they make their honeycombs from various flowers; they build wax cells, and replenish their fortress with innumerable offspring; they have armies and kings; they wage battle; they flee smoke; they are annoyed by disturbance. Many people know from experience that bees are born from the carcasses of oxen, for the flesh of slaughtered calves is beaten to create these bees, so that worms are created [from] the putrid gore, and the worms then become bees. Specifically, the ones called 'bees' originate from oxen, just as hornets come from horses, drones from mules, and wasps from asses.
>
> (Book 11, 4: 3 and Book 12, 8: 1–3)

Such opinions, reminiscent of Aelian, were transcribed in all the books on beekeeping and agriculture, and were in effect until well into the nineteenth century; for example, the *Agriculture Treaties* of Alonso de Herrera (1513): "of some young bull bees are born."

As a second example of the importance of medieval entomology, and much more in line with modern science, we should mention the scholastic Albertus Magnus. Born in Lauingen, Bavaria and died in Cologne (1193–1280), he was a priest, bishop, and Church doctor. He was a prominent theologian, geographer, astronomer, philosopher, botanist, chemist, and naturalist with great vocation as experimental observer and representative figure of chemistry and, in general, a polymath of medieval science (Strunz 1925; Aiken 1947; Weisheipl 1980; Castillo 1996; Lima 2009). In his empirical and encyclopedic work, he combined Platonic doctrines with Aristotelian ones, and faithfully ensured them to be compatible with Christian doctrines. His ideas were thought in European universities until all Aristotelian doctrines were condemned in 1210 by the Council of Paris. However, he exceeded Aristotle with his own observations and contemporary data, especially in Botany and Chemistry.

Beside his first work *De anima*, dedicated to the Natural World, we must cite in particular his magnum opus *Parva Naturalia*, the most remarkable part of which is Book XXII, *De animalibus* (*c.* 1256–1268), which classifies species based on their properties. In Books VIII and XXIII he devotes attention to dissection and study of the nervous system of some insects, crabs, and scorpions; observes the loss of orientation in ants with amputated antennas and deduces that their eyes must be on their antennas; and makes detailed observation of the methods used in the construction of cobwebs. With vocation and zeal, he conducts experiments to study various arthropods: he details his experiments of immersing scorpions in different fluids (*De animalibus* 26, 33: 39) or the insensitivity to fire in spiders (*De animalibus* 25, 35–6: 46). In the twenty-six volumes of *De animalibus*, he credits all the medieval entomological knowledge and his interpretations to the hand of God who created them. The first nineteen volumes are devoted to the works of Aristotle which he helped disseminate, the next two to his own contributions, and the last five correspond to the *Books of Zoology* (*De natura rerum*) by Thomas Cantimpré (Cantiprantanus), and include writings of Avicenna and Galen, emphasizing the importance of observation and scientific rigor. Although few species of insects are covered (of the 450 species mentioned only thirty-three are insects), he nevertheless admires their morphological perfection and makes numerous observations. He dismisses or corrects myths and demystifies legendary entomological elements, although curiously he notes of flies having eight legs (Aiken 1947).

Other major entomological works during the thirteenth century include books of Alexander Neckam (*De naturis rerum, c.* 1200); Thomas Cantiprantanus (*Liber de Natura Rerum*, 1233–1248) with a treatise on the life of bees (*Bonum universale de apibus*), a moralistic and didactic work, which compares the life of a monastic community with the beehive; Bartholomeus Anglicus (*De proprietatibus rerum*, mid-thirteenth century); Vincent of Beauvais (*Speculum Universale/ Majus*), with *Speculum naturale* Books XVIII–XXII dealing with wild or domestic animals such as dogs, snakes, bees, and insects; and Juan Gil de Zamora (*c.* 1241–*c.* 1318) and his *Liber contra venena et animalia venenosa*. These books also contained drastically incorrect statements about insects, butterflies, and some "worms and flying scorpions," and perpetuated Aristotle's theory of spontaneous generation (living beings arising from inanimate matter) which remained in Europe until the middle of seventeenth century (Hays 1973; Morge 1973; Cloudsley-Thompson 1976; Crombie 1990, 2003; Beullens 2007; Page 2007). This was commonplace in the Classical World even before Aristotle: the genealogical, ethnographic, and geographical materials written by the *logographoi* (ghost-writers) such as Scylax of Caryanda, Charon of Lampsacus, or Hecataeus of Miletus relied heavily on oral and written traditions before their time, and as Thucydides (1.21) pointed out, "the logographoi ... were more interested in entertaining their audience than in telling them the truth."

New discoveries during the Middle Ages in Europe, new entomological data provided by travelers and adventurers contributed to the effervescent medieval zoology and entomology. Manuscripts and diaries of travelers such as Benjamin of Tudela (1130–1173), William of Rubruck (1220–1293), and

Marco Polo (1254–1325) were widely circulated and translated into many languages and provided new information and new iconic animals unknown in the West (Page 2007).

We should not forget that during the Middle Ages, animals were commonly used to convey moral lessons, and each animal was assigned an instantiated, exemplified, and educational mission. The *Physiologus* and Bestiaries are excellent examples (Robinson 1961; McCulloch 1962; Clair 1967; Barber and Riches 1971; Clébert 1971; Baur 1974; Mode 1975; Mermier 1977; Curley 1979; Bianciotto 1980; Henderson 1982; Diekstra 1985; Yapp 1985; Schrader 1986; Payne 1990; Sorabji 1993; Houwen 1997; Malaxecheverría 1999; Ruskin 2000; Campagne and Campagne 2005; Beullens 2007; Resl 2007a, 2007b; Zambon 2010; Heck and Cordonnier 2012, etc.). Different species of animals were marked with a didactic and moralizing significance (Klingender 1971; Roberts 1982; Shaver Crandell 1989; Sorabji 1993; Chevalier and Gheerbrant 1997; Carmona Muela 1998; Steel et al. 1999; Camille 2000; Werness 2006; Beullens 2007; Castelli 2007; De Leemans and Klemm 2007; Page 2007, etc.). Naturally, the popularity and authority of these texts—whether classical, sacred, or from travelers—encouraged the fantastical medieval imagination and increased their zoological richness to a point of no limits and no return, where real and fantastic beings became part of a credible and truthful reality. Images of unicorns in Creation scenes or on Noah's Ark were not uncommon.

In this medieval imagery is where we find a greater number of arthropod references, although one should not expect to find butterflies, grasshoppers, or dragonflies as we now understand, or to be able to differentiate clearly between a scorpion and a crab or a lizard under our modern concept. Apart from the "real" beings, one should consider that the Middle Ages is the age of the bestiary, of monsters and chimaeric and fantastic beings that characterize it (Schrader 1986; Benton 1992, 2002; Hassig 1995; Mezzalira 2002; Mittman 2006; Page 2007; Beullens 2007, etc.). These beings are intrinsic elements to the imagery of the Middle Ages and proliferate in medieval iconography and architecture, especially in the Romanesque (Figures 5.1–5.6), for almost a thousand years (Mode 1975; Friedmann 1980; Guerra 1986; Shaver Crandell 1989; Hicks 1993; Davy 1996; Flores 1996; Chevalier and Gheerbrant 1997; Houwen 1997; Camille 2000; Réau 2000; Ruskin 2000; Monserrat 2011c, etc.). Harpies, locusts, basilisks, faucets, and other monstrous beings were heavily represented in a wide variety of interpretations and aspects (Figures 5.1–5.6). They often possess a tail (of a serpent, dragon, or scorpion), an element introduced to the Christianity from the Mesopotamian and Hindu legends, namely the Neo-Assyrian deity *Pazuzu* (Black and Green 1992; Roaf 2000) or the Mesopotamian *girtablullû*, made up of the head of a human, body of a bird, and tail of a scorpion. These appear with some frequency in the Kasite, Neo-Babylonian, Neo-Assyrian, and Seleucid glyphs (Buren 1937–1939). The chimeras are also mentioned in the books of the *Iliad* and the *Aeneid* and also described by Hesiod in Teogonia. The Ethiopian manticores quoted by Ctesias from Pliny (VIII, 30) are reminiscent of the devastating locusts with destructive scorpion tails from Revelations (Figures 5.1–5.6). Medieval arthropod elements, especially beings with scorpion tails, are also present in zodiac signs (Figures 5.8–5.12).

FIGURE 5.8 External capitals of the Ducal Palace (1344) with the horoscope of Cancer and Scorpio, Venice (Italy). Photograph by Victor L. Monserrat.

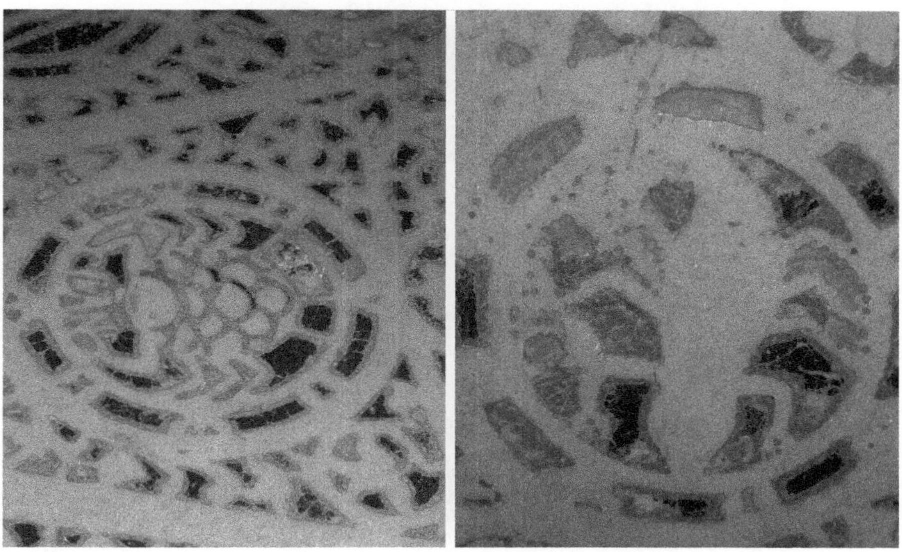

FIGURE 5.9 Zodiacal elements in the pavement of the Baptistery of Santa Maria dei Fiore (eleventh century), Florence (Italy). Photograph by Victor L. Monserrat.

FIGURE 5.10 Romanic archivolts of the central door of San Marco (1508) with motifs of the year's activities and symbols of Cancer and Scorpio, Venice (Italy). Photograph by Victor L. Monserrat.

FIGURE 5.11 Zodiacal elements in the Clock Tower (1496), Piazza San Marco, Venice (Italy). Photograph by Victor L. Monserrat.

FIGURE 5.12 Zodiacal elements in the main left door, Church of Notre Dame, Paris (France), in the four squares in the top left is Cancer, the crab, and Scorpio is in bottom square. Photograph by Victor L. Monserrat.

The character of animals was considered unambiguous and consistent, while that of a man was ambiguous and variable, and thus served the moralizing purpose of symbolizing virtues and defects, elements of nature, the cosmos, etc., not to mention as symbols of antagonism and domination of some over others. Animals (and insects) were historically subject to all symbols of good and evil, and in the fearful Middle Ages, full of ignorance and supernatural superstitions, increasingly populated the lists provided in *Physiologus* and Bestiaries. The inclusion of insects and other animals in the known medieval Books of Hours and Breviaries is not only to serve as decorative motifs, but their presence in these beautiful miniatures is loaded with a vast moral and instructive symbolism (Cambefort 2004; Beullens 2007; Page 2007; Resl 2007a, 2007b; Monserrat 2016).

Unlike today, when we have immediate access to a wide range of documents and visual information on any subject, the zoological and entomological medieval world had no source of information other than the imagination of artists, which was in itself based on the often-confusing information received from the above texts and their iconography.

As with animals, some insects were also taken by their characteristics and behavior as symbolic elements. Ants have been repeatedly cited by classical authors such as Aristotle, Herodotus, Aelian, Pliny, etc. They are treated in Proverbs (6:6) in the words of Solomon, as well as in relation to the parable of the Five virgins and their oil lamps (Matthew 25:8). St. Augustine considers them wise, farsighted, and industrious, and decent for man to imitate (Enarratia in Psalmum lxvi and Prov.vi, xxx, 25). Ants also appear in the fourteenth-century Italian bestiary, *Fiori di Virtu*. In Christian symbolism, ants are always associated with the virtue of prudence, and are an excellent example of the moral interpretation of animals used by classical authors (Werness 2007). In *Physiologus* (chapter 55) and the Bestiaries, they are extolled for their qualities in relation to the virtues and spirituality of Christians, and are referenced in anti-Semitic judgment against the Jewish people who, as they thought then, "lack these virtues" (IV, 1–4) (Bulard 1934, 1935). It was said that the ants

> ... walk in order as soldiers; They carry grains in their mouths, and a grainless ant does not attempt to take the grain from another; they break each grain in half to save germination; when it rains they get the grain out to dry, they collect grain in summer to not starve in the winter; when it is time to harvest the grain, they enter the fields and go up to the grain, where they distinguish wheat from barley by their smell, and reject barley because it is food for cattle.
>
> (Medieval Bestiaries)

Ant-lions have been cited by Megasthenes, Herodotus, and in Ovid's *Metamorphoses* with some mythical propensities. The Romans coined the term *myrmekoleon*, meaning an ant-faced feline. In certain medieval texts, ant-lions appear either as lions or carnivores (such as the Anglo-Saxon *De Rebus in East Mirabilibus*), and in some others (such as some French Bestiaries) they appear as ants. Examples of the latter are the *Bestiary of Philip of Thaun* (no. 8), the *Bestiary of Guillaume le Clerc of Normandy* (no. 11), the *Bestiary of Gervaise* (CAP. 16), and the *Bestiary of Pierre de Beauvais* (long version no. 28). These were probably following Job 4:11, who refers to ant-lions (mistranslated as "lions") in a quote from Eliphaz, King of the Temanites, pointing out the struggle with starvation due to their dual identities: the lion-faced father who eats meat and the ant-faced mother who eats grains, and therefore the ant-lions can't consume either meat or vegetables. Similar to the ant-lion, man can't follow two paths but only one: that of salvation. This dual monster is sometimes cited as *formicaleon*, as in a fifteenth-century Bestiary based on *De Proprietatibus Rerum* (*c.* 1240) of

Bartholomaeus Anglicus, and is recorded to have eight legs. Pope Gregory the Great (540–604) made reference to ant-lions' successful carnivorous larvae buried in the sand, but stigmatized it as demonic and associated with hypocrisy. St. Augustine also associated it with the devil. More scientifically, St. Isidore of Seville (*Etymologies*, 12, 3:10) maintained the common name of ant-lion, but resolved the myth by interpreting them as "lion of ants," a character that he associated with Myrmeleontidae larvae (Neuroptera) who bury themselves in the sand and hunt ants with their big jaws. Other works, such as *Hierozoicon* (1663) of Samuel Bochart (Chapter IV), *De Universe* of Rabanus Maurus (*c.* 844), and *Bestiis et aliis Rebus* of Hugo of Fouilloy devoted attention to this mythical insect (Gerhardt 1965; Kevan 1992; Monserrat 2011b) which they called *formicaleon* and aptly described the trait of living in the sand: "... it is a small animal enormously hostile to the ants. It hides in the dust and kills ants carrying their supplies" (Rabanus Maurus, *De Universe*).

Many other examples that could be cited on medieval entomology, in some cases, are downright "fun." For example, Pierre de Beauvais in his *Bestiary* (1218) says, "the cricket likes to sing so much that it forgets everything, loses its appetite, lets itself be hunted, and dies singing." A similar idea is presented in the Cambrai Bestiary (thirteenth century): "while the cricket sings, he forgets to eat and dies singing."

To all this, we must add legislative and ecclesiastic elements specific to the medieval entomological mentality of the Church. In the Middle Ages, exorcisms, exhortations, and prayers were frequent, and animal trials were not uncommon. For example, it was believed that caterpillars were sent by God to punish sinners and were subjected to excommunication. St. Bernard started the trend of excommunicating some annoying insects, in this case flies that pestered him. The bishop of Lausanne in 1478 led the excommunication of *Melolontha* larvae affecting their crops. The archbishop of Leon in 1520 dictated the excommunication against the caterpillars and locusts that did terrible damage in the harvest by the clergy of Troyes in 1516 (cited in the Archives of the cathedral of Leon, Abbaye Saint-Loup de Troyes, and Abbaye Saint-Martin-ès-Aires de Troyes). There are numerous instances where the intercession of the Pope was requested to alleviate invasions of insects in various locations, and many votive offerings painted or carved on wood that were used in the nineteenth century to request the intervention of God in such cases (Lea 1883; Zarco-Cuevas 1997; Pastoureau 2006). One of the first "insecticides" used in the West was the holy water, with which areas invaded by locusts, caterpillars, and flies were sprayed. St. Gratus, bishop of Aosta in the fifth century (and the Patron Saint of the Fear of Insects) used this method to eliminate an invasion of weevils and other insects, which were regarded as divine punishment and were attempted to remove by processions, payment of tithes, confession, and fasting (Ratcliffe 2006).

The Judeo-Christian arrogance of man over the animals regarded them as beings lacking intelligence and sensitivity, a notion supported by later thinkers and philosophers such as René Descartes (1595–1650), who argued for their uncontrolled and irrational behavior. Hundreds of animals, including insects, were judged if they could feel pain, because if they did, they were considered "consciously bad" or possessed, and even judged and punished by execution, cursing or excommunication (Pastoureau 2006). Some examples include: the punishment imposed in 1541 on locusts in Italy; the public performance of a sow infanticide which occurred in Falaise (France) in 1386; or the lawsuit initiated by the Abbey Párraces (Spain) for the extermination of the locusts in 1650. Such practices continued until 1765 (Lea 1883; Zarco-Cuevas 1997; Pastoureau 2006). The habit of excommunicating insects even came to the New World, and has been documented as late as 1713, when the Franciscans of the province of Piedade in Brazil denounced the ants that had invaded their monastery. The Franciscans represented ants by a lawyer, and demanded that the ants assume and obey the judge's sentence: they should leave the territory free of occupation of ants, and they should move with haste from the monastery to the underground or risk the severe penalty of excommunication. A friar read the sentence to ants and "insects loyally accepted it and headed to their new home" (Lea 1883).

The secular counter to this irrationality and cruelty to animals emerged through the works of philosophers like English Jeremy Bentham, who proposed that animals could not think, but they did suffer. This vision, however, was already voiced by some critical thinkers among theologians around 1648 and until the eighteenth century. Defending the rights of animal resulted in the creation of the British Society for the Prevention of Cruelty to animals in 1824, which to this day with numerous associations and organizations, continues to fight for protection, respect, and defense of animals.

Today, in our urbanite and sanitized western civilization, our contact with nature is almost always accidental, and our relationship with animals and insects is mostly anecdotal and casual. But in the Middle Ages this was not the case: insects and other arthropods were part of man's daily life, and in a world much more in touch with nature and their avatars, during the Middle Ages insects were an essential part of rituals, liturgies, beliefs, and actions, including those of the Christian saints.

We have already discussed insects that are cited in the Revealed texts, but examples also exist in the pious and fearful Middle Ages with its long Christian tradition of sainthood, following the entomological tradition of Greco-Roman mythology (Waddell 1946; Zarin 2006; Page 2007). Evidence of this can be found in St. Ambrose (340–397), bishop of Milan (374–397) and one of the Big Four Doctors of the Church, who baptized St. Augustine. The etymology of his name comes from ragweed (*Ambrosia*), as immortality nectar reserved for the gods, goddesses, heroes, and their horses and angels. Although it was not permitted for mortal Greeks (and punishment of Tantalus is an example), with permission from the gods, some mortals (e.g., Aristeo, son of Apollo) could rise to a superhuman status. Ragweed was associated with the Eucharist and the Word of God, which gave immortality to believers. St. Ambrose, born in Trier and raised in Rome, patron of stonecutters and protector of bees, and patron of chandlers and beekeepers, tells in *Golden Legend* of how bees fly "in and out of his mouth, as if they were making honey," depositing the honey of the theological knowledge on his lips, while he slept in the cradle in the courtyard of his family home (Vorágine 2014). This is apparently plagiarized from pagan Greeks, who used the same story about the poet Pindar (called "sweet as honey") for his eloquence, or the Latin Eliano (nicknamed "honey tongue") for his mastery of Greek. Other classical associations, such as that between the eagle and Jupiter, will evolve into the association between the eagle and Christ, and the Delphic connotations between the snail and resurrection will also be adopted in the story of the resurrection of Christ.

Some Egyptian sources were also co-opted by Christianity. Such is the case of the deeply rooted symbolism of the sacred Egyptian scarab (Ward 1902; Weiss 1927; Swift 1931; Cambefort 1987, 1994a, 1994b; Werness 2007; Monserrat 2013a) which was later associated with Christ by Christian Saints, such as St. Ambrose of Milan, St. Augustine, and St. Cyril of Alexandria. More curious still is the association of bees ("Flies") with the persecuted and perverse practice of idolatry: the term *Pope fly* (Spanish: *Papa moscas*; from the common name of a bird *Ficedula hypoleuca*, "Flycatcher") was used to define those who practiced idolatry, and is used to refer to a naïve, candid, or gullible person. The term appears in some medieval texts such as *Le livre du pèlerin de vie humaine* (1355) of Guillaume de Deguileville (Edwards and Maupeu 2015), as well as in statues and clocks of medieval cathedrals, such as the *Papamoscas* of the Cathedral of Burgos.

Bees were associated with the weapons of the Egyptian sun god Ra and were the intermediaries between the gods and humanity (Ransome 1937; Kritsky 1993; Kritsky and Cherry 2000; Monserrat 2013a). The relationship of the Egyptian god Apis with idolatry for Christians and consequently with the figure of the bull is well known, but also *Apis* (= bee) could explain this ancestral link between the figure of the bee and the concept and medieval Christian image on the idolatry.

Among the lesser-known Christian saints is St. Bernard of Clairvaux (1090–1153), a Cistercian monk who associated the bees with the Holy Spirit and compared the hives with the monastic community of orderly life. He is associated with honeycombs and has the hive as an attribute (he was known in the Middle Ages as "master of the sweet honey," "mellific doctor," or "honey mouth," and was appreciated for his devotion to the human aspects of Christ). The Greek bishop John Chrysostom (344–407), known for his eloquence ("golden mouth"), was associated with bees because his speech was "sweet as honey," a term possibly plagiarized from pagan Greeks. St. Norbert of Prémontré (1080–1134), a noble German courtier who became archbishop of Magdeburg in 1126, is attributed to the spider by a passage of his life when he drank wine from a chalice, inside of which was a poisonous spider. He then rejected the sacrament, and is represented with a spider on a chalice. A similar story also exists about St. Clement of Alexandria, who based on Proverbs—"go to the ant, O lazy one, observe her ways and be wise" (6:6–8)—cited ants as an example of prudence and foresight against laziness, and is often depicted with these insects as attributes.

Many other Catholic or Orthodox saints have insect stories associated with them, such as St. Theophilus of Philadelphia, who recommends "Imitate the wisdom of bees" and cites them as an example of monastic spiritual life; St. Basil who uses the metaphor of the butterfly and the Resurrection; St. Colman, bishop of Lindisfarne (*c.* 610) whom Gregory the Great compares with bees (Letter VII, 24, 1, 16–20) and who fled from worldly riches accompanied by humble creatures like the rooster, mouse, and fly, and a fly served him as flag to mark pages in his readings; St. Sossima (probably derived from the ancient Russian god of bees *Zosim*); St. Macarius who was stung by flies in desert; St. Leutfred (738) with flies that would not let him nap; St. Gobneta, who was in charge of the Convent of Ballyvourney in York (Ireland) until the year 500, and used bees against the bands of outlaws that terrorized villagers; or St. Narcissus, bishop of Gerona in the fourth century, associated with flies due to the prodigious fact that occurred in 1286 when invading French troops tried to desecrate his grave. Similar reference is made to bishop of Nisibis in the year 346 who used mosquitoes in hostilities between the Romans and Persian King Shapur. Flies, bees, spiders, and their webs are also associated with other saints such as St. Felix of Nola, St. Conrad of Constance, St. Sabbatius (Sawatij), etc.

With this information basis on the origins of medieval entomological thought within Christianity, it is time to move to the next chapter where we will address the significance of insects as symbols during the Middle Ages.

CHAPTER SIX

Insects as Symbols

VÍCTOR J. MONSERRAT AND VAZRICK NAZARI

In the previous chapter we analyzed the origins of medieval entomological thinking within Christianity. We will now discuss additional aspects of the symbolism and significance of different orders of insects and arthropods during the Middle Ages from a non-religious standpoint. We will examine available medieval sources, focusing on some of the entomological elements that have persisted in the folklore, traditions, and legends in some European communities.

As mentioned in the previous chapter, unlike the civilizations of antiquity (e.g., Mesopotamia, Mycenae, Egypt, the Greco-Roman world, etc.), where arthropods were a substantial part of people's beliefs and mythologies and often came to acquire the status of deities (e.g., cicadas of Minerva, bees of Zeus, grasshoppers of Apollo, the Egyptian sacred scarab Khepri, etc.), in Europe, after the rise of Christianity and throughout the Middle Ages, insects and other arthropods were mostly relegated to the dark underworld of the demonic, sin, and evil. We have inherited this mentality in the form of various insect-phobias as well as an undeserved "bad reputation" for arthropods (Swift 1931; Tod 1939; Davies and Kathirithamby 1954; Morge 1973; Cambefort 1987, 1994a, 1994b; Mellinkoff 2004; Ratcliffe 2006; Monserrat 2011b, 2012a, 2012b, 2013a, 2013b, 2013c, 2014, 2016b).

From our contemporary perspective, the medieval profusion of evil and moral dangers strongly disregarded the beautiful and unspoiled natural world, and morality was extended to be represented by animals (Sorabji 1993). Although we only focus on insects and some other arthropods, it is evident that they are mostly associated with vice, sin, death, and darkness, as opposed to a very small representation of nicer attributes and allegories (virtue, purity, industriousness, hierarchy, monastic community, or order). Insects are almost never associated with good, beauty, and light. Poor and illiterate people lived in a sad, dark, and unhappy medieval Europe, mired in superstition, uncertainty, fear, and punishment. All of these had effects on the symbolism, traditions, and popular beliefs of insects in the Middle Ages.

A wealth of literature exists on zoological or entomological symbolism in medieval and Christian Europe (Berry 1929; Frost 1937; Brion 1959; Murga 1983; Cantó Rubio 1985; Hicks 1993; Roy 1994; Flores 1996; Chevalier and Gheerbrant 1997; Cirlot 1997; Carmona Muela 1998, 2000; Grant 1999; Kitchell and Resnick 1999; Bruce-Mitford 2000; Camille 2000; Cooper 2000; Resl 2007a, 2007b). It is important to remember that our current Cartesian conceptions of insects and arthropods does not always correspond with the medieval ideology and classification. For example, the medieval image of the scorpion presented in the Bestiaries and Zodiacs may be indistinguishable from our current concept of a crustaceans, amphibians, or reptiles (McCulloch 1962; Klingender 1971; Hays 1973; Mermier 1977; Bianciotto 1980; Roberts 1982; Diekstra 1985; Schrader 1986; Clark et al. 1989; Hassig 1995; Malaxecheverría 1999; Clark 2006; Zambon 2010; Dyer 2013). The signs of Cancer and Scorpio on the pediment of *San Isidoro de Leon*, Spain (s. XII) are difficult to differentiate (Figure 6.1) (Monserrat 2011c). During the Middle Ages, insects were mixed in a hodgepodge together with lizards, geckos, certain arthropods, and a variety of other "bugs," both as a concept and in iconography. In an analysis of medieval iconography, Ruskin (2000) placed scorpions in the same category as reptiles. Indeed, for the medieval observer, these were the same thing: bugs with legs that wag their tail. The medieval mentality was more concerned with whether or not otters or beavers were fish, and if they could or couldn't be consumed during Lent. One must therefore keep an open mind when interpreting the (often contradicting and confusing) entomological elements in medieval ideology and iconography (Benton 1992, 2002).

Apart from personal interpretations of texts by medieval theologians and polymaths over the centuries, most of these inconsistencies come from trying to adopt and adapt the pre-existing European folklore and Greco-Roman heritage with the new Christian ideology. In such instances, the cultural significance of many animals changed, sometimes drastically to the opposite (Panofsky 1955; Evans 1969; Child 1971; Klingender 1971; Morge 1973; Ettlinger 1978; Scarborough 1979; Payne 1990; Charbonneau-Lassay 1996; Steel et al. 1999; Yanoviak 2013, etc.). Such is the case of ancient symbolism of the sacred Egyptian scarab (Cambefort 1987, 1994a, 1994b; Monserrat 2013b, 2014) which, through the works of St. Ambrose of Milan, St. Augustine, and St. Cyril of Alexandria, was symbolically associated with Christ. However, it was linked with sin, evil, and temptation, presumably by pagans (Sprecher and Taroni 2004; Monserrat 2016a). Other examples include the strong associations

FIGURE 6.1 Zodiacal elements with signs of Cancer and Scorpio in the pediment of San Isidoro (twelfth century), León (Spain). Courtesy of Nuria Cantón de Celis.

between Jupiter's eagle and Christ, or the Delphic connotations of the snail and the Virgin Mary or the resurrection of Christ, and between Egyptian mongooses who devour the evil snakes with the Christ figure.

However, some translations influenced by the Christian religion are purposely done erroneously, in order to avoid pagan connotations. Such is the case of the owl of Minerva (or the owl of Athena), an animal that represented wisdom, shifted its definition and came to be associated with heresy, cruelty, loneliness, darkness of the night, and death. In fact, the owl received more than twenty-five possible symbolic interpretations over the centuries, from wisdom to stupidity, and from epilepsy to the Jewish people. As another example, in *Moralia*, Pope Gregory the Great presented the dreaded swarms of locusts as a symbol of *Conversa gentilitas*, comparing them with the pagans surrounding Christ who revolted as a swarm to attack Satan. Another marked shift in symbolic meaning that contradicted ancient concepts as well as the sacred writings was that of the turtle, considered an abominable animal in Leviticus 11:2–47 but which later came to represent Christ's sepulcher, chastity, modesty, silence, and wisdom of the elderly. Such surprising contradictions were often introduced in translations of the sacred texts from original sources to confront what was considered pagan influences by biased translators (Ferguson 1959; Child 1971; Ettlinger 1978; Grabar 1980; Guerra 1986; Blasco Zumeta 1997; Houwen 1997; Revilla 2003; Monserrat 2014, 2016a, etc.).

As an example of such disparity of interpretations in entomological literature, Gagliardi (1976) described seventy-four different symbolic contexts in which butterflies can be cited or appear. Part of these contradictions may be justified on the basis that certain body parts or aspects of a particular animal have been used for purposes of very different interpretations. All those variables would make the analysis difficult, but we will try to organize all possibilities and find the most likely interpretation in each context.

Taking in consideration all these contradictions and leaving aside all fantastic animals (i.e., griffins, sirens, manticores, harpies, etc.) and real non-insect creatures (mammals, amphibians, reptiles, fishes, birds, non-arthropod invertebrates, etc.), we will only focus on the symbolic meanings associated with "bugs" in the medieval era.

SUMMARY OF ARTHROPOD SYMBOLISM IN THE MIDDLE AGES

Ant: tenacity, constancy, prudence, foresight, spiritual qualities (which the Jewish people lack for Christians), agitation of the clergy at the service of the Church, greed, covetousness.

Ant-lion: duality, devil, demonic, hypocrisy. Man cannot follow two paths but only one; i.e., that of salvation.

Bee: chastity, obedience, good community, monastic life, order, purity, courage, prudence, wisdom, Jesus, virginity, Virgin Mary, resurrection, soul, word of God, organization, hierarchy (Figure 6.2).

Beehive: motherhood, hope, monastic community, orderly life, family, monarchy (Figure 6.2).

Beetle: immortality, resurrection, Christ, demon, temptation, sin, sinners (Figure 6.3).

Butterfly: lightness, inconstancy, soul (both of sinners and condemned as of righteous and saved), resurrection, salvation, virginity, purity, modesty, Virgin Mary (Figure 6.3).

Caterpillar/chrysalis: body of the Savior, resurrection, human body, spiritual retreat.

FIGURE 6.2 Detail of a miniature of bees guarding their hives against a marauding bear, from Flore de virtu e de costumi (Flowers of Virtue and of Custom), Italy (Padua?), second quarter of the fifteenth century. © British Library Board, Harley MS 3448, f. 10v.

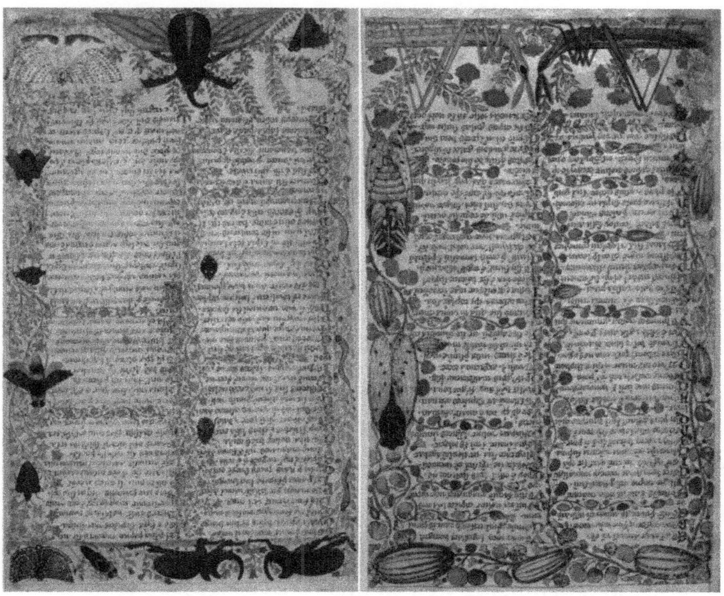

FIGURE 6.3 Cocharelli Book, leaves from a prose treatise on the *Seven Vices* and a verse on events from the history of Sicily in the time of Frederick II (1298–1337), Italy, (Genoa), *c.* 1330–*c.* 1340. © British Library Board, Add. 28841, f. 4v, f. 5v. Along the top margin of the left page is a beetle, along the bottom margin of the left page on the far right is a butterfly, and two beetles are on the right side of the bottom margin. On the right page, top margin, are two locusts, and two cicadas are along the center margin of the right page.

Cicada: alternation, duality, neglect, hedonism, unpredictability, music, sense of hearing, hope, resurrection, summer (Figure 6.3).

Crab: invulnerability of Christ, resurrection, greed, deceit, sin, inconstancy, shrewdness, bad neighborhood, cruelty, temptation, vice, Jewish people (for Christians).

Cricket: surveillance, bliss, home.

Dragonfly/damselfly: sin, evil, demon.

Firefly: The light of Christ.

Fly: devil, torment, Passion of Christ, evil, pain, precariousness of life, perishable, transient earthly life, shamelessness, ignominy, carnal desires, futility, pain.

Honey: sweetness of Christ, Eucharist, asceticism, purification, wisdom, oratory, suffering.

Lice or fleas: pests and diseases.

Lobster: resurrected Christ, laziness.

Locust/grasshopper: temperance, heathen who join faith, destruction, devil, irresponsibility, unpredictability, smallness before God, scourge of God, armies, discord, indiscipline, envy, famine.

Mantis: prayer, orientation, good guidance, Christ, counselor.

Mayfly: the brevity of life, the life of Jesus.

Moth: death, ignorance, temptation, sin.

Scorpion: evil, betrayal, disaster, darkness, falsehood, hypocrisy, heresy, deceit, betrayal, impostors, murmurers, perfidy, fornication, paganism, devil, death (synagogue, Judas Iscariot, or Jewish people for Christians), envy, hate, greed, torment, logic, fortune, wisdom, eloquence, dialectics.

Silk: excess, indecency, lust, sin (and "licentious" women).

Spider: esoterism, devil, deceit, Judas, dialectics, ability, greed, lack of firmness, temptation, sense of touch, rain, heterodox doctrines.

Stag beetle: evil, devil, sin, defense against all kinds of evils.

Wasp: heresy, disorder, confusion, ill-treatment, evil.

Wax: initiation, disciple seeking wisdom, spirit of Christ, soul, inconstancy, infidelity, falsehood, belonging to a strain.

Worm (fly larva): dirt, rot, death, sin, soul of sinners, infernal spirit, guilty conscience, remorse, Christ.

Now let us develop these elements group by group:

INSECTS

Hymenoptera

Among the Hymenoptera represented in some medieval illuminated manuscripts, bestiaries, fables, etc. are bees, bumblebees (*Bombus*, and *Xylocopa* to a lesser extent), wasps, ichneumons, and ants (Monserrat 2016a). Bees have had a very long presence in human evolution and have influenced all the cultures with which they have come into contact. The honey bee is the best-treated insect in medieval ideology, symbolism, and iconography, and even though the bee is one of the few common insects that sting, there are almost no negative connotations associated with it. This again goes to show the absurdity and inconsistency in medieval and western entomophobia (Edwards 1908, 1920; Ransome 1937; Fraser 1951; Hallman 1951; Neufeld 1978; Marchenay 1979; Crane 1983, 1999; Benton 1992, 2002; Monserrat 2011a, 2011b, 2012a, 2012b, 2013a, 2013b, 2013c, etc.).

Christianity inherited the false idea from the Greco-Romans world that beehives (Figure 6.1) had a king (not a queen) and that bees reproduced asexually, which led to their association with virginity and the Virgin Mary. The notion of industrious and obedient *virgin bees* has a literary tradition that dates back to Aristotle, Plutarch, and Aelian among the Greek authors and Virgil and Pliny among the Romans.

Plutarch cites them as "irritable and bellicose with men who have been with women" and "trying to attack women who have had sex," although Aelian qualifies, limiting them to "licentious" relationships. In medieval artwork, honey bees and beehives sometimes appear in Nativity scenes with Mary and the child reflecting the symbolism related to virginity, such as the eleventh-century *Exultet Roll* in the Cathedral of Gaeta (Italy). Bees are also associated with female figures and scenes related to purity and virginity against vices. This is evident in the *Psalter of Queen Elizabeth of England* of the Cathedral of Ely, where a feminine figure appears next to a beehive, while the men are drinking and gambling (Monserrat 2016a).

Many ancient superstitions have sprung up around the bee, including a central European tradition instructing a bride to walk her future husband past a beehive to test his fidelity: if he is stung, it is a sign that he will not be a faithful husband. The symbolic relationship between honey bees and virginity is also reflected in the wedding ceremonies around the world, as the term *Honeymoon (Luna de miel, Luna di miele, Honigmond*) can testify.

In the Proverbs, bees are instantiated as role models. Clement of Alexandria calls *Sicilian bee* one of his teachers. An old proverb among Christians said "the same way among bees there is a King, the prince of prophets is my Lord

Jesus Christ." Such comparisons are frequent in the Church's history even in modern times: in 1958 Pope Pius XII considered bees "a proof of God's existence."

Virgil mentions bees as an example of order and divine intelligence; he compares their castes with the Christian virtues and their morphology with the essence of man, with a spiritual side that resides in the wings, a carnal carrying the sting, and the narrow link between the thorax and abdomen which is comparable to the fragile equilibrium between the human body and soul.

St. Tertullian compared the life of Christ with bees, and Pedro de Padua called the flight of the soul towards the heaven *Apis aetherea*. Thomas of Cantimpré (*c.* 1259) compared the observance of monastic life with the chastity of bees, and mentioned that the monks should follow the life of the bees like the ones of St. Sturm, founder of the *Monastery of Fulda*. Beehives have been used as symbols for the ideal human society by many authors, from Aristotle to Tolstoy. The straw beehive (skep) became a symbol of the Papal miter during the Renaissance. Because of the structure of their societies, their helpful role in nature and their attractive product, in Medieval Ages bees were associated with productivity, industriousness, and creativity. Even today they are incorporated in the logos of many companies, banks, cities, or associations.

Another example of medieval entomological misconceptions inherited from classical authors (i.e., Antigonus, Virgil, Eumelus, Columela, Aelian, Archelos, Celsus, Florentinus, Varro, Ovid, Pliny, *Book of Judges XIV*, *Geoponica*, etc.) was the belief that bees were generated from cattle corpses, an error that linked bees with death and resurrection. This may have been a result of confusion between bees (Hymenoptera: Apidae) and *Eristalis* flies (Diptera: Syrphidae) (Osten-Sacken 1894; Atkins 1948; Freire García 1995; Barbattini and Nicoli Aldini 2010, etc.). This belief persisted until well into the nineteenth century, and were transcribed to treaties of agriculture and apiculture, for example the *Agriculture Treaty* of Alonso de Herrera (1513): "of some steers are made bees."

The link between bees and resurrection in Christianity is also rooted in ancient Greece. During the three months of winter, the activity of bees seems to subside, before they are seen again in the first sunny days of spring. Christians somehow associated this period with the three days during which Jesus Christ was resurrected after his death. A Gelasin sacrament mentions the bees and their skills "they suck the flowers without damaging them ... as well, thanks to their lips, they become mothers, the same way as Jesus came from the lips of God."

In Mesopotamian, Egyptian, and Classical cultures, bees were intermediaries between gods and mortals. Bees and their products (wax and honey) entered the realm of death and the rituals of embalming of the dead. This is also still preserved in rituals, liturgies, sacraments, and ointments within the Christian religion and mythology. St. Ambrose is usually considered the patron of stonecutters and chandlers but also of beekeepers, and the protector of bees and hives. His image appears in the medieval iconography represented with bees, hives, or honeycombs. Even today in Bulgaria on February 10 people gather to honor St. Haralampi, the protector of beekeepers. The profession of beekeeping in the West has left a multitude of references in languages, place names, surnames, and even items of clothing (Friedmann 1980; Réau 1997a, 1997b, 1998, 1999, 2000).

In Britain and Ireland a strong tradition of bee folklore persists. One medieval superstition states that if a bee flies around your house or buzzes at your window, it means that a visitor will soon arrive, and if someone kills the bee, the visitor will bring nothing but bad news. In Greek folklore, if a bee lands on your head, it is said to mean that you will be very successful in life, and if a bee touches the lips of a child, they will grow up to be a wonderful poet.

Bees are also believed to be very sensitive creatures, and in Britain they must be spoken to politely and informed of all the family news (indeed, if you wish to rid yourself of bees, the quickest way is to swear at them as they despise bad language). In medieval Germany, in a practice known as "telling the bees," bees were informed of the death of the owner of the hive with a wand. Sometimes straw beehives were draped with a black cape before telling the bees that the beekeeper had died. The tradition of "telling the bees" varies regionally, but the most important information to impart to the bees is when their owner dies, and the bees must be sensitively told of the death or they will desert the hive, cease making honey, or die. In New England, where bees were introduced around 1638, bees were informed of the death of a family member in order to prevent them from leaving the hive. In some cases in Britain and America, treating bees as part of the family became so well integrated that they would be invited to family weddings or funerals and given a piece of the wedding cake.

Neuroptera (Myrmeleontidae)

Ant-lion larvae with their peculiar hunting behavior are one of the oldest mythological insects (Druce 1923; Gerhardt 1965; Kevan 1992; Monserrat 2010a, 2016a). They were thought to be part ant and part lion (*myrmecoleon*), or a "lion among ants" which hid in the dust and killed ants. This puzzling duality led ant-lions to be seen as symbols of the dual nature of the human soul: part heavenly (ant) and part demonic (lion), and so they were associated with hypocrisy.

Hemiptera

Various true bugs (Hemiptera, Pentatomidae: *Carpocoris fuscispinus, Palomena prasina, Rhaphigaster nebula, Graphosoma*; Notonectidae: *Notonecta glauca*; Pyrrhocoridae: *Pyrrhocoris apterus*; Acanthosomidae, etc.) have been illustrated in medieval illuminated manuscripts as marginal decorations, embellishment (Figure 6.2), or in some symbolic context linked to the Virgin Mary (Monserrat 2016a). Cicadas (Cicadidae) also have a very long tradition in the Classical world, but are free from any divine attributes. They were associated in the Middle Ages with alternation, duality, negligence, hedonism, or improvidence, and in better connotations, music, the sense of hearing, hope, resurrection, or summer (Figure 6.2) (Monserrat 2011b, 2012a, 2016b). Cicadas' association with good luck has made it a very popular insect in the traditions of some regions of Europe, especially in Mediterranean countries. In the south of France, images of cicadas are used in various objects of family use. Cicadas' sound has been a source of inspiration in music and the making of numerous popular instruments shared throughout the Indo-European continent: the Hindu *Toulouhou*, the *ratsche*, *ratchet*, *rattle*, *sonaglio*, or *hochet* in Germany, England, France, and Italy, and the *carraca* in Spain.

Coleoptera

The strong Egyptian influence in Coleopteran symbolism is well known (Swift 1931; Cambefort 1987, 1994a, 1994b; Ratcliffe 2006; Monserrat 2013b), but their presence in the symbolism and traditions of the West goes back to the Greeks and Romans, when many customs associated with these insects existed (Davies and Kathirithamby 1954; Monserrat 2011b, 2012a, 2013a, 2013b). In Greco-Roman mythology, we find stories linked to Zeus such as the shepherd Cerambus who was transformed into a wood-gnawing beetle (*Cerambyx*).

The beetle is one of the larger and more contradictory entomological elements of medieval symbolisms (Benton 1992, 2002). As we mentioned in the previous chapter, reminiscent of the Egyptian world, St. Ambrose of Milan, St. Augustine, or St. Cyril of Alexandria symbolically associated the beetle with Christ, and as a consequence, beetles became emblems of immortality, resurrection, and defense against all kinds of evil. However, coleopterans (especially the stag beetle, *Lucanus cervus*; Coleoptera: Lucanidae) also came to represent sin and sinners, evil and the devil (Monserrat 2016a).

In the Middle Ages, there was a predominant dislike and persecution of insects similar to other animals such as cats, bats, and nocturnal birds of prey. People believed that insects carried danger, evil, sin, or misfortune.

This absurdity of such a persecution has remained until today in the form of the vulgar and derogatory term "bug" to refer to any type of arthropod even the harmless ones. The stag beetle, one of the most persecuted and punished insects, has a large number of legends and representations (Impelluso 2003; Sprecher and Taroni 2004). Beside its negative connotations, as is normal throughout the iconography of the Middle Ages, this beautiful and particular beetle was also linked to the triumph of God over evil, and was a symbol of defense against all kinds of dangers. It is relatively abundant in medieval miniatures and other artistic representations (Sprecher and Taroni 2004; Monserrat 2016a). Stag beetle-related traditions were long held in ancient Europe (Crete, Greece, Rome, etc.), and continued even after the fall of the Roman Empire, especially among the Gnostics in Sardinia and east of Italy where beetles and snakes were the only animals accepted for sacrificial practices. The use of these stag beetles in children's games and as pets (possibly a predecessor of kite games in Europe) was common among the Greeks and then the Romans (Davies and Kathirithamby 1954; Monserrat 2013c), as mentioned by Antonius Liberalis (*Metamorphosis XXII*). Romans severed the beetles' heads and made necklaces for their children to prevent illness. Remnants of such practices exist today in Bavaria, in the Vosges in eastern France, and in Austria, where the mandibles of stag beetles are used as medicine against cramps.

Medieval legends also hold that stag beetles protected humans by making sure that lightning strikes them and not the people, perhaps because they were found on old oak trees that were struck by lightning. Among the Germans, the god Thor had a link to *Lucanus cervus*, and this beetle was greatly feared in the Nordic cultures because they believed it could extend fires by moving the embers and making the cattle run away. The stag beetle was used, along with other arthropod elements, in heraldry (Fox-Davies 1969; Dennys 1975; Woold 1996).

Apart from the stag beetle, several other beetles were associated with religious figures and were omens of good luck. The firefly (*Lampyris noctiluca*, Insecta, Coleoptera: Lampyridae) was a symbol of the light of Christ. Two others, namely the ladybug (*Coccinella septempunctata*) and the rose chafer and flower beetles (*Cetonia* spp, *Potosia* spp.) (Insecta, Coleoptera: Coccinellidae, Cetoniidae) had a significant presence in popular folklore, and had (and still have) a certain relationship with the divine and the sacred. Folktales in Anjou or Guyenne (France) state that rose chafers bring happiness during the eighth *Corpus Christi*, especially if they are on roses (*Rosa canina*) gathered to decorate the processions.

The much more popular ladybugs are most often associated with the Virgin Mary. The seven black spots (i.e., *Coccinella "septempunctata"*) were associated with the seven sorrows of the Virgin (Yanoviak 2013). Perhaps the reputation of the mystical and Kabbalistic number seven played a role in the ladybug's popularity: many folkloric traditions relate to counting the number of spots on a ladybug's back which may reveal how many children, or how many months of good luck, or how much money one should be expecting. In English folklore, it is said that if a ladybug lands on your hand, you will be married within the year.

A popular medieval legend tells of a terrible swarm of insects (probably aphids) that descended upon the fields in Europe, destroying precious crops. People implored the Virgin Mary for help, and supposedly she sent an army of red and black beetles to consume the invaders. Since then, they've been also called "Our Lady's Bugs," and ultimately "Ladybugs." Since then, ladybugs are seen as a sign of divine protection. If a farmer's crops were being devastated by aphids, they would pray for ladybugs, who would come and eat the aphids—thus saving the crops.

In France, the ladybug is called *bête à bon Dieu* or *bête du bon Dieu* (beast of good god); in Piamonte, it is called *pollo de san Miguel* (chicken of St. Michel). In Tuscany, it is called *Lucia* and, in England, *ladybird* or *ladyfly*. In Sicily, it is called *palumedda* and children invoke it when they bury a deciduous fallen tooth, getting a reward after. In Germany, it is called *God's little bird* or *Cockalorum of Mary* and it is invoked when people want to get milk and butter, so it is present in the verses of the homeworkers. In Spain is called *Mariquita*, *Bicho de la Virgen* o *Vaca de Dios* (Virgin's bug or God's cow) and is present in children's songbooks ("Mariquita, quita, quita, ponte el velo y vete a misa": Ladybird, get away, get away, put on the veil and go to Mass). It is also present in other countries such as Russia, Germany, France, or Italy. Particularly in Switzerland, children ask the ladybug how many years they will live.

Ladybugs are also associated with renewal from medieval times. It has been thought that a ladybug landing on old clothes might be indicative that the clothes will soon be replaced, and that a sick person might find a ladybug flying away with their illness, gifting them with a renewal of health.

References to other beetles are abundant in European traditions and beliefs. One example is the dung beetle *Geotrupes stercolarius* (Geotrupidae), which in Finland is granted powers to defend the crops from the storms; in Austria it is associated with ghosts; in Sweden and Germany it is associated with witches and supernatural powers; and in France, legends hold that this beetle drank the last blood of Jesus on the cross (it is known that when provoked, this beetle exudes a drop of red fluid).

Insects and other arthropods are present throughout the world of medieval legends, beliefs, superstitions, and sorcery (Weiss 1912, 1930; Wyndham 1960; Gubernatis 1968; Siganos 1985; Pollock 2005; Lawrence-Mathers and Escobar-Vargas 2014, etc.), and legacies of entomological witchcraft were abundant in medieval medicine (Walsh 1920; Getz 1998; Jones 1998; Porter 1999; Cambefort 2004; Mitchell 2004; Bowers 2007; Wallis 2010; Siraisi 2012; Khateeb Shehada 2013, etc.), and reached the Enlightenment (Benito del Caño and Roldán y Guerrero 1928; Monserrat 2013d).

In the Early Middle Ages, following the fall of the western Roman Empire, standard medical knowledge was based chiefly upon surviving Greek and Roman texts, preserved in monasteries and elsewhere. Many simply placed their hopes on God and the Church to heal their sicknesses. Ideas about the origin and cure of disease were not purely secular, but were also based on a world view in which factors such as destiny, sin, and astral influences played as great a part as any physical cause. The efficacy of cures was similarly bound in the beliefs of patient and doctor rather than empirical evidence, so that *remedia physicalia* (physical remedies) were often subordinate to spiritual intervention.

In relation to medieval medicine and these *remedia physicalia*, we have many references to the use of minerals, herbals, and botanical remedies, but also parts of animals, including insects. The extensive medieval pharmacology maintained its Roman entomological heritage in the West until the birth of modern pharmacology in 1900. Until the nineteenth century, medicinal prescriptions included animals such as crabs, scorpions, centipedes (*Scolopendra* sp.), bees, beetles, cantharids, etc., as well as insect derivatives, such as bee propolis, honey, jellies, waxes, centipede water, ground bee dust, scorpion oil, beetle oil, crab eye dust, cantharid pills, mealybug powder, homopteran syrup, etc. (Getz 1998; Porter 1999; Cambefort 2004; Mitchell 2004; Bowers 2007; Siraisi 2012).

In medieval Europe, the "medicinal" use of beetles was widespread, and apothecaries used them in many remedies (Benito del Caño and Roldán y Guerrero 1928; Monserrat 2013d). The case of *Lytta vesicatoria* (Coleoptera: Meloidae), popularly known as cantharides or "Spanish fly" (which is neither fly nor Spanish only), is well known. Since the time of Greeks and Romans, it was used in medicine as a vesicant for skin ulcerations, alopecia (hair loss), urinary incontinence, and also as a diuretic, while sometimes creating priapism as a secondary effect. Cantharids were very popular as an aphrodisiac for centuries, and were frequently used by sovereigns such as Fernando el Católico, who is said that to have died due to abuse of the "Spanish fly." Other royals or nobilities also used it for the same purpose: Henry IV, members of the Medici family (who called it *Aqua Toffana* or *Aquetta di Napoli*), Louis XIV, and, of course, the lascivious Marquis de Sade were amongst them. This insect was especially popular in Europe throughout the Baroque and Enlightenment as shown in the illustration *Pastilles Richelieu*, but it fell into disuse due to the high number of poisonings (Monserrat 2009b, 2013c). Even today there are websites offering cantharids as aphrodisiacs.

Diptera

We have already discussed in the previous chapter the dipteran as an element present in all cultures and civilizations, from Mesopotamia to Christianity. The link between flies and death is a biological reality documented since antiquity.

In the Medieval Period, we have reference to beliefs associating the fly with torment, the passion of Christ, evil, pain, the precariousness of life, perishable things, shamelessness, ignominy, lust, and futility. Both flies and the "worm" (fly larvae) are linked to dirt, decay, death, sin, temptation, deceit, the souls of sinners, the infernal spirit, consciousness guilt, remorse, and with Christ, the liberator of all these evils (Pigler 1964; Chastel 1984). With such a wide range of symbolism, the fly appears very frequently in medieval illuminated manuscripts (Cambefort 2004; Monserrat 2011c, 2016a).

Other dipteran, such as mosquitoes, do not have such quaint associations as some other insects but are almost universally perceived as a menace due to their annoyance and nasty bites. It is therefore no surprise that most superstitions around mosquitoes relate to ways of preventing them from biting. One such superstition is that if you eat some green vegetables on Maundy Thursday (also known as Green Thursday), then mosquitoes will not bite you for an entire year. An old wives' tale also says that if you make your bed on new hay during the harvest time, the mosquitoes will not bite you (Cock-Starkey 2016).

Orthoptera

Within this order we mentioned the cricket, a common insect in many European traditions, and a symbol of vigilance, bliss, home, and family during the Middle Ages (Figure 6.2). These symbolic characteristics have remained in many European customs. Such is the case of the cricket cages that are still sold today in the "Cricket Festival" in Florence, Italy, held on the day of the Ascension (coinciding with the ancient May calends of the Romans). Another example of similar entomological folk craft in Europe is pottery made in the town of Villafranca de los Caballeros in Toledo, Spain, in the form of spherical cages with a door to introduce the insect into the cage, and to provide him with food and fresh air (Pelauzy and Catalá Roca 1977). In some parts of Europe, the ancient habit of having one or two crickets at home for good fortune has been maintained for centuries, although in areas of Celtic influence, the song of the cricket inside the houses presages a death.

Another European tradition related to crickets is preserved in Santa Maria de Huerta in Soria, Spain, where male crickets are collected from the field to compete in a contest where the winner is the cricket with the loudest song. An infantile tradition related to crickets existed in Villafranca de los Caballeros in Toledo and some other areas in La Mancha, Spain where it was customary for the village grooms on their wedding night to go out and hunt crickets by performing a "manhood contest" that relied on their "power of urination." They urinated in the bushes where crickets were thought to hide, forcing them to leave. They would then capture and bring the poor cricket back as a trophy of their "masculinity" (Seseña 1997).

Isidore of Seville (*c.* 556–636) mentions crickets and crickets cages: "This animal walks backward, bores through the earth, and chirps at night. It may be caught by tossing an ant tied with a string into its hole, first making a puff of dust so that it doesn't hide itself. When the cricket grabs the ant, it is drawn out of its hole." Isidore's account seems to be based on a tradition in Asturias (Spain), where children hunt crickets by introducing a straw or ants into his gallery, which forces the cricket to come out (Mariño Ferro 1996).

There are many poems and riddles among the farmers in France, Italy, Spain, and Portugal related to these singing insects, which are very common in the Mediterranean region. In Asturias, Spain, black crickets are called black coals and yellow crickets are kings: "I am an eagle without being a bird, without being a king I have a crown," "I am black without being charcoal, day and night I play the harp, and without being a king I wear a crown and the river I cross without a boat." In Portugal, crickets are considered the king of the winged insects, and in the Beira Alta province in northern Portugal, they are bringers of good fortune. On the contrary, the locust was an evil creature used in witchcraft and exorcism for the Romans, and an emblem of Apollo for the Greeks. Instances of the former practice are also mentioned in the Sacred texts cited in the previous chapter: Christians maintained many of those folkloric traditions, relating them to destruction, famine, and other negative connotations. Evil locusts are mentioned numerous times in the Apocalypse of St. John, and the spread of his works in the Middle Ages through *Los Beatos* translations which had a tremendous negative effect on the perception of insects in Christianity (Monserrat 2014).

Ephemeroptera

Mayflies were used to symbolize the brevity of life, as well as the life of Jesus. Due to their use as fishing bait, they were a very popular insect among European fishermen and their stories. Mayflies appear infrequently in medieval iconography, but are depicted among the marginalia in several illuminated manuscripts (Monserrat 2016a).

Odonata

Dragonflies and damselflies are considered to have positive values in the East, while, in the West, the opposite is true. Indeed, they are associated with danger, bad omens, sin, evil, and the devil. As such, their image is frequent in medieval iconography, especially in illuminated manuscripts (Monserrat 2016a).

According to Forrest and Lasswell (2005), an old Romanian tradition says that when God sent St. George and an army of angels to put the Devil in his place, St. George's horse was possessed by the devil and turned into a flying insect. Not surprisingly, the Romanian term for dragonfly is *calul dracului*, meaning "devil's horse." The idea of dragonflies and damselflies as the devil's tool pervades European folklore and is reflected in the many names colloquially given to these creatures; i.e., the English words dragonfly, devil's needles, the Spanish *caballito del diablo* (devil's little horse), and the French *aiguille du diable* (the needle of the devil). In German they've been given a number of folkloric names, including *Teufelspferd* (devil's horse) and *Wasserhexe* (water witch), whereas in Danish they were known as *Fandens ridehest* (devil's riding horse). Swedish tradition also associates these winged insects with the devil, as they are reported to weigh the souls of people.

In southern England as well as the US, odonates are still associated with superstitions related to snakes and fishing. They are said to be able to differentiate good people from bad; it is considered an unlucky omen to kill one or if one bites you. In American folklore, dragonflies were thought to be "snake doctors," since the two creatures are often seen together. It was believed that if a snake was cut in half, the dragonfly would use its long and thin body to sew the reptile back together. In English folklore, damselflies were known as "the devil's knitting (or darning) needle," because it was believed that if you went to sleep next to a stream, the damselflies would use their long bodies to sew your eyelids shut (Cock-Starkey 2016). In France, it was believed that dragonflies could spit poison into people's eyes, and if a person was bitten by them, the affected area should be amputated. In Sweden, it was thought that a dragonfly could pick out a person's eyes, and in Old Swedish these insects were called *blindsticka* ("blind stinger"). In Ireland, Scotland, and Italy, they were considered witches who lived in the forests and could sew the mouths of lying children. None of these beliefs, however, seem to have existed before Christianity; for example, in Germanic mythology damselflies were associated with Freya, the goddess of poetry, fertility, love, and eroticism, perhaps because when two damselflies mate, their bodies appear together in the shape of a heart. Astonishingly, a new ordinance in the year 775 changed the latter beliefs and officially linked these insects with paganism and evil.

Dermaptera

Despite popular belief, earwigs really don't want to live in your external ear canal to lay eggs, even though they do like to live in dark and wet places. Their common name is possibly the oldest name of any insect in English language, dating as far back as the eleventh century. "Earwig" derives from the Old English *ear wicga*, which roughly translates to "ear insect" or "ear wiggler" (*wicga* meaning "wiggle").

Dictyoptera

There are very few insects that have few or no negative labels in medieval times. One of them is the mantis, associated with praying, orientation, good guide, and a counselor to Christ.

A major part of classical Greek mythology involved divination (*manteia*). Ancient Greeks believed the mantis to be a spirit sent by Hecate from the most sinister of hells. Mantids are present in many Greek fables and tales (Henderson 1982; Bell 1992), often associated with Hecate, the goddess of wealth, wisdom, navigation, magic, sorcery, divination, and crossroads. Mantids were also associated with the Greek underworld (Avernus) and Princess Cassandra. The Romans believed that evil fell upon any animal or man who looked at a mantis's eyes, as its gaze was considered to cause ailments of unknown origin, or it was said to those who were sick, "the mantis must have looked at you." The divinatory mantis, inherited from the Classical world, retained its presence in Christian mythology, but lost the maleficent magical powers that it had during the Greek times (Davies and Kathirithamby 1954; Schimitschek 1968, 1974; Monserrat 2009a, 2010c, 2011b, 2012a, 2016b) (see Dante's *Divine Comedy*). Instead, it became associated with prayer and worship, as evident in the scientific names of the most common species in Europe (*Mantis religiosa* and *Iris oratoria*) (Dictyoptera, Mantodea).

Some legends and myths surrounding the mantis have to do with finding direction, such as the medieval belief that mantids guide travelers and children who are lost in the woods. Early French folktales say that a lost child could find their way home by following the directions of a praying mantis. Intriguingly, there is an old Arabic folktale among Muslims that tells of the mantis pointing towards Mecca. This finding-of-direction ability sometimes expanded to include both physical and emotional directions, and mantids became omens of good guidance in general. Later, during the Renaissance and Baroque era, it was discovered that the female mantis sometimes eats her male partner after they mate, and thus mantids acquired a new fame as symbols of sexual misogyny and aggressive sexual power.

The medieval Christian Church excluded real animals from a dominant place in Christianity, and it was forbidden to worship them and to use their bodies in rituals of communication with God (Page 2007). However, in the superstitious Middle Ages, magic and witchcraft were often practiced without persecution. The art of identifying and interpreting meaningful patterns and signs in nature, including the sounds, movements, and appearances of animals, parts of their bodies, or predictions from the flights of birds or bees, were maintained in medieval beliefs as divinatory systems (Page 2007; Lawrence-Mathers and Escobar-Vargas 2014).

During the Late Middle Ages, supernatural assistance was often sought for curing sick animals, for attracting desirable animals (such as bees and birds), and repelling vermin and predators like mice, snakes, and wolves. The means to access these supernatural benefits varied from those generally viewed as licit—prayers, charms, blessings, and curses—and those classed as illicit—magic, superstition, or idolatry. Medieval people drew on the powers of patron saints of animals (see the previous chapter), celestial influences, or words and images that were thought to have power over particular species (Page 2007; Lawrence-Mathers and Escobar-Vargas 2014). People thought that animals could act as channels of supernatural power in the physical world, helping saints or punishing sinners according to God's command (Page 2007).

Lepidoptera

In medieval imagery, butterflies symbolize lightness and inconstancy, but with Classical Greek influences, especially that of the soul. They also represent salvation, purity, modesty, and virginity. Moths, on the other hand, represent death, ignorance, temptation, or sin. The caterpillar and chrysalis symbolize the body of the Savior, resurrection, and the spiritual retreat.

Butterflies appear in many medieval texts and images in association with the Virgin Mary (Gagliardi 1976, 1996, 1997; Domínguez Rodríguez 1993; Verdon 2005; Monserrat 2016a). Among the most common elements enhancing the idea of virginity, innocence, and purity in medieval artwork is the presence of a white butterfly, very often white pierids (Pieridae: *Pieris* sp.). The earliest example of this is from Irish monks of the Order of St. Gall (*c.* 800–1000), who included white butterflies, among bees and honey, in verses dedicated to the Virgin Mary. This symbolism remained in many art forms even after the Renaissance.

Being one of the most beautiful insects, butterflies have been the subject of numerous allegorical connections. We discussed how Gagliardi (1976) described seventy-four different symbolic contexts in which butterflies may appear, and among them are some of the most interesting contradictions in the entomological medieval mentality. We have already mentioned the relation with the soul, the Virgin Mary, and the virginity of maidens, and also that sin and evil has to be fought with and scared away (maybe because moths are nocturnal animals that fly like owls, bats, and other creatures considered "evil"). This is why butterflies frequently appear in medieval manuscripts associated with human figures, where a knight (symbol of good) fights with a butterfly or moth (symbol of evil), using a sword, spears, shields, arrows, or other weapons. On several occasions, the villain butterfly has bat-like wings or it is a bat-winged butterfly with antennae which might suggest the link between them in the medieval mentality. In illuminated manuscripts, butterflies (symbols of evil or souls of sinners) appear in many scenes being chased either by centaurs, cherubs (*putti*), monkeys, or people, sometimes with hoods or bare hands (Monserrat 2016a).

In the Greco-Roman myth of *Cupid and Psyche*, best told in the novel *The Golden Ass* (or Metamorphoses) by the Roman fabulist Apuleius (*c.* 125–*c.* 180), Psyche is an earthly woman whose beauty threatens Venus, the goddess of Love. Venus sends Cupid to take revenge, but Cupid falls in love with her. Venus banishes Psyche to the underworld but she comes back to life victoriously, and is granted immortality by Jupiter so that she can marry Cupid as an equal. Hence, she symbolizes not only the image of the immortal soul, but the anguish and triumph of soul. The Greeks and Romans saw butterflies as personification of Psyche's death and resurrection cycle. In Greco-Roman artifacts featuring scenes with Cupid and Psyche embracing or otherwise engaged in various amorous or entertaining activities, Cupid is always depicted with angelic, feathered wings like those of a bird, while Psyche has fragile, often highly stylized wings similar to those of a butterfly (Davies and Kathirithamby 1954; Monserrat 2011b, 2012a; Nazari 2014).

This myth survived in the symbology adapted by Christianity, and is frequently depicted in medieval artistic representations relating to souls or judgment (Blatchford 1889; Brewer and Sandved 1976; Gagliardi 1976, 1996, 1997; Charbonneau-Lassay 1996). The scenes of the trial of the tympanum on the west door of the Cathedral of Autun (St. Lazare) in France (*c.* 1130–1135) are an excellent example of a human figure with butterfly wings. Later, in late medieval or Renaissance tables or Baroque paintings, a butterfly was often placed in the hands of the infant Jesus as allegory of the soul.

The image of the caterpillar or the chrysalis among Christians is thus strongly connected with the notion of eternity and resurrection of souls, since it is a reference to the human body of Jesus, and therefore a positive element. Among the Greeks there was a strong association between the butterfly and the soul, the body, and any atavistic sense. The presence of an association between butterflies and death and rebirth in the mythology and folklore of a wide variety of native peoples around the world suggests a very old human tradition. Similar stories can be found among pre-Columbian native Americans like the Goajiros of Colombia, the Aymara of Bolivia, and between Mayas and Aztecs, as well as among Native Americans, peoples of the Pacific, Africa, and Asia, such as Maori, Australians, Japanese, Samoans, Dayakos, Ibans of Sarawak, Burmese, Solomon Islands, Yakuto and Chuvacho, Nagas, Zambezi Ila, Kasai Balubas, Central African Lulu, and Bakongos, etc. In many of those cultures, hunting or causing harm to butterflies is considered taboo. Butterflies are often part of their rituals and rites related to death and resurrection, and sometimes considered as the spirits of the dead. Among the Mahorais of New Zealand, butterflies represent immortality and the soul; in Iran and Anatolia there is a belief that the souls of the deceased appear in the form of moths and the deceased can appear in the form of nocturnal butterflies. There are also numerous Mayan and Aztec citations on this phenomenon (Heyden and Baus Czitrom 1991), perhaps best exemplified by the monarch butterflies (Danaidae, *Danaus plexippus*) who arrive in Michoacan on the Day of the Saints, after an eight-week migration and a 2,000–3,000 km flight from North America. Monarch butterflies are considered to be the souls of their dead by the Aztecs.

Some atavistic elements of classical origin remain in European folklore; for example it is considered bad luck to handle a hairy caterpillar, because touching one can leave nasty spines that can cause irritation and lesions in one's hand. It also is said that the bad luck can be negated by tossing the (poor) caterpillars over one's left shoulder.

These lepidopteran beliefs, which were also prevalent among pre-Roman (or non-Romanized) Europeans, such as the Celts, Saxons, Germans, Eskimos, etc., can be found even in modern Europe. For example, in Finland it is believed that during a dream, the soul of people travels in the form of a butterfly, which would explain the memories that remain with us when waking up (Grinnell 1899; Hogue 1975, 1987, 2009; Brewer and Sandved 1976; Gagliardi 1976, 1996, 1997; Berlo 1983; Bentley 1988; Heyden and Baus Czitrom 1991; Costa Neto 2000; Headrick 2003; Monserrat 2010b, 2011a, etc.). The association between lepidopterans and death is also reflected in species that carry the name, for example, the death's-head hawkmoth *Acherontia atropos* (Lepidoptera: Sphingidae), a species that is intimately linked to superstition and bad omens in Europe from Celtic times (Klingender 1971; Duval 1977; Kliffen 1996), with contemporary cinematography (*The Silence of the Lambs*, Jonathan Demme, 1991). Black or dark butterflies, especially species of *Saturnia* (Saturnidae) or *Erebia* (Satyridae), are still very much linked to vices and bad omens in certain parts of Europe; for example, in Westphalia, Germany, on February 22 they still celebrate a ritual day for the expulsion of moths and nocturnal butterflies in order to avoid bad luck. In Scotland, Friesland, Serbia, and Bosnia, lepidopterans are considered witches: a Serbian proverb says that "if you kill a butterfly, you kill a witch." For the Celts, the presence of butterflies at night around houses was not a good sign, and villagers and farmers saw them as signs of upcoming hunger or fatality.

In Sicily, the butterfly is called "the little bird of good news" and is related to good luck. In the west of Scotland, for a period of time, they used to eat butterflies with honey because of its divine symbolism. They were linked to the soul and fire, since they were attracted to it. Even the Scottish word for butterfly (*dealan-de* = brightness of God) and one of the names for butterfly in Gaelic (*teine-de* = fire of God) were used in ceremonies related to the maintenance of fire. The English "butterfly" (Anglo-Saxon "buttor fleage or buter flege") or German ("Schmeterling, Molkendieb") seem to come from the white color of the common cabbage butterfly (*Pieris brassicae* or *Pieris rapae*) and for being the first butterfly flying in springtime, when butter or cream was being made.

The European languages leave us a lot of historical information. For example, the Irish and Breton word *Maro*: (goddess of) death; the Lithuanian *Moré*: goddess of death or old witch; the Greek, Germanic, and Slavic *Mora*, *Mara*, and *Morava* meaning both nightmare and butterfly; and other words of German, French, or Castilian, such as *Mahr*, *Cauchemar*, or *Mariposa*, may have this ancestral root, although in the case of *Mariposa*, Christianization of the "pagan" Europe forced the origins and intentions to be linked to Virgin Mary (it lands on Mary).

OTHERS ARTHROPODS AND CRUSTACEANS

Scorpion

The totemic scorpion is one of the ancient arthropods in the history of our species from its earliest Neolithic manifestations (Göbekli Tepe, *c.* 11,600 BP; Çatal Hüyük, *c.* 7,400–6,200 BP; Samarra, fifth millennium BCE, etc.) and one of the most cited in mythology and legends of many cultures around the world (Bulard 1935; Buren 1937, 1939; Tod 1939; Hall 1974; Cloudsley-Thompson 1976, 1990, 2001; Gimbutas 1991, 1996; Mariño Ferro 1996; Moser 1998; Kritsky and Cherry 2000; Monserrat 2009c, 2010b, 2012a, 2012b, 2013b; Monserrat and Melic 2012, etc.). This chelicerate appears in medieval Christianity with symbolic elements that date back to antiquity: the scorpion was linked to malevolent men, Tertullian adopted it as an emblem for the heretics, and it was a permanent element in Mitraic temples. The German theologian Rabanus Maurus (*c.* 776–856) was the first to associate scorpions with the devil. Christianity associated scorpions with treachery and perfidy, and consequently to the Jewish people, even though Jesus was Jewish himself (Lucas 3:23–38; Mathew 1:2–16). Scorpions often appear in the iconography of the last days of his life. They are prominently represented in European shields, robes, banners, and artworks between the fourteenth and sixteenth centuries (Monserrat 2009a, 2009c, 2010c).

The scorpion is one of the most perverse arthropods for Christianity. In medieval Europe it was linked to many evil and negative attributes; it personified Judas Iscariot or the Jewish people for Christians. However, later it also became linked to logic, fortune, wisdom, eloquence, and dialectic. Even though the scorpion is a Zodiac sign, its image does not possess any of the aforementioned meanings. There are numerous representations of scorpios (or crabs) in medieval architecture, reliefs, religious and decorative objects, or texts related to astrology, astronomy, or medicine that reflect the aesthetic of the medieval models, sometimes looking too plain, and sometimes with meticulous details showing the personality of the artist or his vision of nature. Some examples include the activities of the month by Benedetto Antelami in the Baptistery of the Cathedral of Parma (*c.* 1220), the reliefs of the Cathedral of Vezelay (XII), of Notre-Dame de Chartres, the Cathedral of Notre-Dame in Amiens (*c.* 1250), St. Austremoine d'Issoire or St. Lazare in Autun, Notre Dame of Paris in France, the porch capitals of Palazzo Ducale in Venice (fourteenth–fifteenth centuries), in the frescoes and pediment of San Isidoro de León (XII) in Spain, and many others. Sometimes the images of scorpios are so deformed that it is difficult to differentiate them from the crab; for example in the front tympan of St. Isidore of Leon (twelfth century) (Figure 6.3) (Monserrat 2011c, 2016a).

Spider

Spiders are also one of the oldest represented arthropods (Çatal Hüyük *c.* 7400–6200 BCE), with a strong ancestral atavism in many cultures which links them to the creation of the cosmos and women (Grinnell 1899; Hall 1974; Gimbutas 1991, 1996; Benton 1992, 2002; Mari.o Ferro 1996; Moser 1998; Kritsky and Cherry 2000; Cloudsley-Thompson 2001; Monserrat 2012b; Monserrat and Melic 2012, etc.). In medieval Europe, the spider started losing its feminine symbolism since female gods were in disuse. The Greek tendency of masculinization of the gods was also maintained by Christianity, and hence the spider also lost its sexual and mythological significance, although remnants of it still remain: in Corsica and Sardinia, it symbolizes sex, disease, madness, and death; in Greece, it is associated with menstruation; and in Italy, there is the belief that the bites of a spider can be healed by dancing with young women. It is also related to esoteric concepts, the devil, temptation, greed, licentiousness, as well as the dialectic, skill, tact, rain, and heterodox doctrines. A curious symbol associated with the spider is the sense of touch. It is found in frescoes like the one of *The Five Senses* (s. XIV) in Longthorpe Tower in Cambridgeshire (Peterborough), where animals such as the pig (hearing), rooster (sight), monkey (taste), vulture (smell), and spider, including its web (touch) are all ruled by the observant king. In several texts, the spider appears to be related to deception and temptation of the sin. The myth of Arachne is common in many European folktales and fables (Henderson 1982; Bell 1992).

Also probably from this origin is tarantella, a popular dance from southern Italy which originated in the Neapolitan origin and extended into Puglia, Basilicata, Calabria, Abruzzo, Molise, Campania, and Sicily. Although its name most likely derives from the city of Taranto in Apulia, it is popularly believed to be related to tarantula, a common spider in the Mediterranean region (ſLycosidae: *Lycosa*, wolf spider). This magic-religious dance is a solo performance, supposedly imitating the mating ritual of the spider. During the Middle Ages, it was believed that dancing tarantella at harvest (summer) time will scare spiders away and cure a type of madness supposedly produced by their bite. The dance was later applied as a supposed cure for delirium and contortions of neurotic women.

Crab

The symbolism of the crab is another ancient legacy from Mesopotamia (Monserrat 2012b). In the Medieval Period, they were related to the invulnerability of Christ, but also linked to greed, deception, sin, inconstancy, bad neighbors, cruelty, temptation, vice, or the Jewish people. Because of its oscillating movement, the crab has been associated with inconstancy and sin. A variant of this animal is the lobster which is also linked to the resurrected Christ and is a sign of sloth.

As mentioned earlier, the crab is often indistinguishable from the zodiac sign of Cancer. In medieval architecture, the crab commonly represented labor (Monserrat 2009a, 2011c, 2016a). Crabs are not uncommon in medieval texts, and may also appear in aquatic scenes in paintings, frescoes, or on the seas of the world maps. Their association with invulnerability makes them appear in heraldic symbols.

Other Cultural and Popular Manifestations Originating from the Symbolism of Insects during the Middle Ages

The arthropod symbolism during the Middle Ages had a remarkable effect on artistic, cultural, and popular manifestations, some of which have survived today. Many aspects of European folklore used in modern European festivals and carnivals—music, dances, expressions, and objects such as masks, horsetails, cowbells, scavengers, etc.—seem to have a medieval root, as they appear to show ancestral connections with cow flies, flies, and other undesirable insects. In the Carnival of Laza in Orense, Spain, participants throw live red ants to the passers-by. In the Carnival of Verín, also in Orense, Don Cigarrón (Mr. Cicada) is a famous personage. Other Spanish examples include the Dantzaris (*Espatadantzaris*: scare-away dancers) in Navarre; the Cachiberrios in the Feast of the Madalena in Briones (La Rioja) which scare the flies towards the *Danzadores* (dancers) before beginning their impressive turns; the scorpions affixed on the layers of the devil's garments in San Antón in Matarraña (Teruel); or the procession to the Hermitage of the Virgin Cows in Fiestas de Santa Teresa in Avila, where live specimens of *Saturnia pyri* (Saturnidae) are released until some settle on the mantle of the virgin, and the clothing and banners of this brotherhood carries this moth as their emblem.

The clouds of locusts and their disastrous consequences for medieval crops have also resulted in traditions that have survived to this day. A traditional folktale from the Lungau Valley in Austria says that the locusts that devoured crops in this area during the Middle Ages only respected daisies and other flowers. As a consequence, the popular custom of imploring the protection of God and his patron saints was initiated by making colorful garlanded poles and constructions made with fresh alpine and meadow flowers collected by the locals from the mountains, including daisies, gentian, peonies, lady's mantle, forget-me-not, poppies, and many more. These flowers are then woven into wreaths and wrapped around the wooden poles (*Prangstange*) to create different color patterns. This centuries-old tradition started because farmers pledged to the Almighty that they would create such floral poles every year if they were spared from similar natural catastrophes in future. The poles are carried in procession through the town twice a year, on June 24 (John the Baptist day) and June 29 (Peter and Paul day). These wooden poles are 6 to 8 meters high and weigh as much as 80 kg. A single pole requires as many as fifty thousand fresh flowers and around three hundred hours of work. The honor of carrying these heavy poles is reserved exclusively for unmarried men from the village.

On the other hand, in a period where the affluent classes formed a hierarchical, chivalric, and ostentatious society, heraldry and its jurisprudence was omnipresent in medieval times (Bartolo di Sassoferrato 1350; Johannes de Bado Aureo 1382–1384; Nicholas Upton *c.* 1446, etc.). The presence of animals, and among them some insects, on the coat of arms of the nobility was not an uncommon phenomenon (Allcock 1962; Keiser 1966; Fox-Davies 1969; Dennys 1975; Woold 1996; Page 2007). The use of images and symbols on coats of arms was initially adopted by military orders around the tenth century and subsequently imposed among the noble classes, giving rise to a complex heraldic system that we carry still today. This tendency for military signs reached a high point in northern France towards the beginning of the twelfth century, mainly due to the need to readily identify the commanders and the soldiers of either side by means of colors and symbols on their armors, standards, and shields in the complex scenes of the battleground. Soon thereafter, figures of birds, animals, and monsters, similar to those seen in the bestiaries, began to appear on shields and banners, which beside identification, were meant to induce intimidation and fear in the enemy (Keiser 1966; Monserrat 2012a). Such figures were also used in medieval jousting and tournaments where, having lost their military significance, they became a complex symbolic system involving identification, social status, and aspirations of lineages, families, and alliances, and turned into emblems of their identity from the blazons of their houses to the sealing of seals of their letters and documents.

After the end of the Crusades, a plethora of decorations, titles, and honors had been assigned, and such accolades were added to coats of arms and other objects that identified each family or personage, from cutlery and table objects to clothes, houses, or monuments as personal and recognizable badges. Many real or fictitious creatures appeared among these identifying elements, especially lions and eagles, often deliberately exaggerated to emphasize qualities appropriate to knighthood (Page 2007). Some insects also received this honor: grasshoppers, crickets, butterflies, ants, bees, wasps, dragonflies, flies, beetles, bedbugs, cicadas, and scorpions, among others appeared very frequently on coats of arms, blazons, and seals that came to identify certain noble, ecclesiastical, or new bourgeois families. In this complex heraldic language, it was habitual in the European blazonry to boast of them in the façades of their seigniorial houses (Jones 1943; Allcock 1962; Keiser 1966; Brewer and Sandved 1976).

Some animals were used to symbolically degrade the status of criminals, heretics, cuckolds, and Jews in punitive and anti-Semitic public rituals and executions, such as the backward ride on an ass, the upside-down hanging with a dog or wolf, or scorpions (Page 2007).

The familiarity, presence, and closeness of insects in the everyday lives of ordinary people during the Middle Ages was much greater than what today, almost anecdotally, is part of our own (Patlagean and Rouche 1991). This has generated many references in legends, tales, stories, and popular sayings developed during the centuries within different countries and languages of Europe (Caro y Cejudo 1792; Bell 1992). To cite an example, Santiago-Álvarez (2012) collected no less than sixty-four common names given in Spanish to different arthropods, and 165 sayings in the Spanish language that include references to arthropods, whether crustaceans, myriapods, arachnids, or insects, of which fifty-one correspond to insects belonging to eleven orders; this is also applicable to many other countries.

Our entomological legacy is not only limited to the Middle Ages. We have inherited a generalized entomophobia that is persistent even today. Hundreds of computer programs and games warn us of the dangers of computer viruses with the icon of a louse, scorpion, spider, or other arthropods; but are they really the culprits?

CHAPTER SEVEN

Insects in Literature and Language

RACHEL MURRAY

There are no insects in Middle English literature; that is to say, the earliest citation of the word in the *Oxford English Dictionary* is Holland's 1601 translation of Pliny the Elder's *Natural History* (79 CE).[1] Strictly speaking, an insect (from the Latin "insectum," meaning "cut up or divided into segments") denotes a class of invertebrate with a tripartite body, three pairs of jointed legs, compound eyes, and a set of antennae (Wigglesworth 2017). As a taxonomic term, it is rather narrowly circumscribed: creatures with too many legs such as arachnids and millipedes are excluded from this category, as are the legless—earthworms, maggots, slugs, and snails. Crucially, though, the writing of the Middle Ages precedes the imposition of such Enlightenment-era taxonomies, in which the animal kingdom is grouped and systematized for the purposes of classification—divided into segments. Instead, insects comprise a somewhat loose order of being that includes worms and spiders, as well as larger organisms such as scorpions and serpents.

Perhaps a more appropriate term for the host of invertebrate life forms that populate medieval literature is the locomotive compound "creepy-crawly," which derives from the Old English verb "creopan," to "move with the body prone and close to the ground, as a short-legged reptile, an insect," and the Middle English verb to "craule," which denotes the act of moving "slowly in a prone position, by dragging the body along close to the ground, as ... any short-limbed quadruped or reptile, an insect, serpent, worm, slug" (*Oxford English Dictionary* 2018). And yet even this designation fails to account for the preponderance of winged critters such as bees and flies that flit and flutter through the pages of medieval texts. And so, for the sake of clarity, it remains necessary to use the term "insect," but in the expanded, popular sense of "comprising, besides the animals scientifically so called, many other arthropods, as spiders, mites, centipedes, wood-lice, etc., and ... earthworms, snails" (*Oxford English Dictionary* 2018). What is becoming clear, however, is just how difficult it is to pin these little critters down in language; throughout human history, and particularly prior to the Linnaean systematization of the natural world, insects have remained somewhat loosely defined, protean entities—partly natural phenomena, partly creatures of the imagination.

As well as being among the most diverse and numerous organisms on the planet, insects feature in a wide variety of literary genres in the Middle Ages from lyric poetry to debates and dream visions, appearing in the form of epic similes, as the subjects of animal fables, and in the margins of manuscripts. These creatures infiltrate the Middle English vernacular as figures of speech, proverbs, and epithets—many of which we still use today. Insects are also highly versatile figures in medieval literature—representative of both the divine and the diabolical, of decay and rebirth, of the joys of collective life as well as the horrors of swarming sociality. Perhaps most disconcertingly, insects can mean everything and nothing to medieval writers: some are valorized for their extraordinary talents, while others are presented as synonymous with qualities of insignificance and negligibility. Consistent, however, is an emphasis on their ability to disrupt and enliven the literary text, generating a swarm of semantic possibilities that pullulate on the page. In his *Natural History*, Pliny notes that "nowhere else does the craft [*artificio*] of nature prevail as in the miniature world of the insect" (Mayhoff 1906: 11.1; Brown 2006: ix). What follows will be an attempt to capture some of the myriad ways that insects inspired the craft of medieval writers across Europe.

FABULAR INSECTS

Throughout the Middle Ages, insects were often apprehended in terms of their ability to reveal essential truths about human nature, as is apparent from the twelfth-century theologian Alain de Lille's pronouncement that "Every creature of the world | Is like a book and a picture | to us and a mirror" (Evans 1983: 151).[2] In addition to larger animals such as lions, foxes, and wolves, insects of various kinds were used as *exempla* in fables to impart moral lessons at the pulpit and in the classroom. Deriving from the Latin *fabula*, or "fiction," a fable is a brief tale that features anthropomorphized creatures of various kinds, and although the structure varies from writer to writer, it usually concludes with an *epimythium* in which the moral of the story is stated explicitly

(Blackham 1985: xi–xix; Wheatley 2000: 3–5; Mann 2009: 29). The fable is closely related to the medieval bestiary, an illustrated book of nature lore that contains descriptions of the physical characteristics and behavior of various animals (both real and imagined), accompanied by didactic material. Medieval bestiaries largely derived from a Greek text known as the *Physiologus*, which circulated through a number of Latin translations. A popular translation of this text that was taught in schools during the Middle Ages features a chapter on the spider, which is depicted as an industrious creature that takes its nourishment from the air (Rendell 1928: 81–3; Curley 2009: xviii).

The majority of these fables were adapted from the *Aesopica*, a collection of stories attributed by Aristotle and Herodotus to a sixth-century Greek slave and storyteller from the Island of Samos known as Aesop (Wheatley 2000: 7–31; Mann 2009: 4–8). These tales are believed to have reached medieval Europe through the Latin prose fables of Romulus and the Latin verse fables of Avianus, before being translated into the vernacular by fabulists such as Marie de France, William Caxton, and Robert Henryson (Lenaghan 1967: 10; Mann 2009: 7–14). The *Aesopica* feature a wide range of anthropomorphic insects, including hornets, ants, flies, grasshoppers, bees, and beetles, and one of the most well-known examples is "The Ant and the Grasshopper." The fable exists in a number of medieval versions, including Marie de France's "Del hulchet e

FIGURE 7.1 Image of a six-legged spider from Jacob Van Maerlant's Der Naturen Bloeme, Flanders, *c.* 1350. Courtesy of Koninklijke Bibliotheek, Ka 16, f. 130r.

de la furmie" [The Ant and the Cricket], Eustache Deschamps's "La fourmi et le criquet" [The Ant and the Cricket], and William Caxton's "The Ant and the Sygale" [The Ant and the Cicada] (Crapelet 1832: 191–2; Jacobs 1889: 123; Spiegel 1987: 127–9). It tells the story of an ant who assiduously stores up grain for winter, while a nearby grasshopper spends the summer music-making. When winter sets in the grasshopper begs for food from the ant, who refuses and instead admonishes the grasshopper for its hedonistic folly.

The tale espouses the virtues of hard work and forward-planning as opposed to the vices of indolence and pleasure-seeking. Caxton's version, for instance, concludes with the maxim: "he that werketh not ne doth no good | shal haue ofte at his teeth grete cold and lacke at his need" [the person that doesn't work or do anything useful | will find himself starving and chattering his teeth in the cold] (Jacobs 1889: 123). Significantly, however, while Caxton's and Deschamp's versions conclude with a straightforward condemnation of the grasshopper, in her early thirteenth-century collection of fables, *Ysopet* [*Little Aesop*], the court writer Marie de France appears more sympathetic towards her stridulating subject:

> "What were you doing," asked the ant,
> 'In August, when you should procure
> Your food and make yourself secure?'
> 'I sing', [the grasshopper] said, 'bring pleasure to
> All creatures, but I find none who
> Will now return the same to me'
>
> (Spiegel 1987: 127)

Although the final lines of the fable are given over to the ant, who chides the grasshopper for his lack of foresight, here the cricket is presented as not merely self-serving, but as a figure who instead "bring[s] pleasure to | All". By presenting the grasshopper as a singer by trade, de France's fable may be warning against the dangers of poetic expression divorced from social responsibility—a sentiment echoed in fellow court writer Richard de Fournival's *Bestiare D'Amours* (1245), the speaker of which states that he has resisted the temptation to write in lyric poetry rather than prose, so as not to be like the cricket that "so delights in singing that it dies in song," adding: "to trust myself to song might mean my self-destruction" (Regalado 2013: 259).

On the other hand, the grasshopper's defense in Marie de France's version of the tale might be read as an attempt to highlight the contribution of the poet to society as a whole, with the author emphasizing that the grasshopper's tireless musical labor is intended to appeal "to | All." Just as Aesop is widely understood to be a figure of low social status whose fables sought to challenge those in more powerful positions, de France's insect fables appear designed not only to reinforce the moral values of medieval society, but also served to disrupt and undermine the status quo (Lefkowitz 2014: 18–20). De France's fable "The Beetle" is adapted from the Aesopic tale of "The Eagle and the Beetle," in which a dung beetle punishes a predatory eagle by destroying its eggs. In de France's version, the beetle is now merely envious of the eagle's flight, and seeks to abandon his lowly dung-heap but finds himself "bewildered, stunned, | And neither could fly higher still | Nor yet go back to his dunghill." The fable highlights the indeterminacy of insects in the medieval imagination, with the beetle lamenting its status as neither "worm nor bird" (Spiegel 1987: 205). The modification of the tale also reflects a shift in cultural attitudes towards dung beetles; in antiquity, these coleopteran creatures were considered sacred, and the Romans, having been influenced by the ancient Egyptian symbolism of the sacred scarab, wore beetle emblems for protection (Zabludoff 2008: 13–17). By the Middle Ages, however, dung beetles had become associated with dirt, filth, and moral decay: thirteenth-century fabulist Odo of Cheriton explicitly associates the beetle with "some who (rather than savoring Christ) savors the dung of sin, the places of the Devil" (Jacobs 1985: 103).

Beetles feature in two other fables by de France, both of which evoke the more abject connotations of coleopteran life forms. In "The Peasant and the Beetle," a beetle crawls into a man's anus, leading him and members of his local community to mistakenly believe that he is pregnant, while "The Wolf and the Beetle" centers on a battle between wolves and insects after a beetle crawls up into a wolf's stomach. Both narratives, which are aligned more with the satirical fabliaux than the moralistic fable, evoke a broader anxiety in the Middle Ages, to be explored in the next section of this chapter, regarding the infiltration of the body by vermin. Significantly, however, in the latter fable in particular, de France complicates the medieval notion of beetles as agents of decay and sin by aligning them with "lesser folk" who join forces in order to overcome oppressive power structures. After crawling in and out of his body in protest, the beetle informs the wolf: "I'm just as good as you are, yes, | ... My people I will gather here | My friends and relations dear | Tomorrow we will fight it out"(Spiegel 1987: 176–7). Seemingly, therefore, "The Wolf and the Beetle" enacts an inversion of the medieval *scala naturae*, or great chain of being, in which smaller creatures such as insects were seen to inhabit a lower order of being than mammalian life forms, and this in turn may be associated with the upturning of feudal power structures (Preece 1999: 121).

After the beetles join forces with wasps, hornets, and bees to sting and invade the bodies of larger beasts, the fable concludes: "The lesser know best how, indeed, | To aid themselves in times of need" (Spiegel 1987: 177). The symbolic connotations of beetles are thus deployed in order to expose and transform hostile social attitudes towards the "lesser" members of society, with de France highlighting the collective ingenuity of the peasantry. It is possible to read this fable alongside "The Ant and the Cricket" as a tacit rebuttal of the poet as little more than a selfish pleasure seeker. In "The Wolf and the Beetle," in particular, de France adapts the insect fable into a mode of incisive, as well as potentially transformative, social critique. The fact that the beetle is taught a lesson in one tale, and teaches a lesson in the other demonstrates that the significance attached to individual insects in the fable, as well as to larger animals, is far from stable: the hero of one tale can be the villain of another. Yet for fabulists such as de France, the inconstancy of insects, which can function as emblems of morality and as figures of dissent, coupled with their indeterminate status as "neither worm nor bird," is perhaps what allows them to challenge rigid thought structures and settled hierarchies. Elsewhere in medieval literature, there are further examples of insects as disruptive figures capable of infiltrating and destabilizing the boundaries of the human body.

WIKKED WORMES

Although de France presents the "worm" as a more definitive category of being than the beetle, throughout the Middle Ages, the Old English term "wyrm," and its Latin equivalent *vermes*, functioned as a catchall term for creeping things of various kinds—maggots, grubs, earthworms—as well as larger organisms such as snakes, lizards, and even dragons (Neville 1999: 108; Thompson 2012: 132).[3] These creatures were believed to generate spontaneously from dead matter, transforming inert substances into teeming life. Isidore of Seville, whose descriptions of natural phenomena in *Etymologies* (600–625 CE) influenced a wide range of medieval writers and thinkers, writes that vermin:

> are generated for the most part from flesh or wood or some earthy substance ... There are vermin of the earth [scorpions, beetles, millipedes, snails], the water [leeches], the air [spiders], flesh [maggots], leaves [silkworms, caterpillars], wood [termites], and clothing [moths].
>
> (Barney et al. 2006: 258)

The entry goes on to identify several kinds of worms that live in the brain, intestines, and skin, and indeed medieval writing is full of accounts of parasitic worms that thrive on the body's natural "liquors" (Ariès 1974: 42; Mellor and Shilling 1997: 39). The Anglo-Saxon medical text, *Bald's Leechbook* (*c.* 800 CE), contains various treatments for different kinds of worm-related ailments: "Læcedomas wiþ þam wyrmum þe innan eglað monnum" [Remedies for worms that afflict man internally], "Læcedomas wiþ wyrmum þe monnes flæsc etaþ" [Remedies for worms that eat a man's flesh] (Meaney 1992: 15). Chaucer's translation of Boethius's *The Consolation of Philosophy* (524 CE) also states that "men wel ofte ben slayn with bytinge of smale flyes" [humans have often been killed by the bites of small flies], he notes, "or elles with the entringe of crepinge wormes in-to the privetees of mannes body" [or by the entry of creeping worms into the private regions of the human body], which suggests that this belief persisted into the Late Middle Ages (Benson 2008: 416).

The frequent emphasis in medieval writing on the infiltration of the body by worms was linked to religious concerns regarding the fate of the body after death: how could resurrection occur in the afterlife if the soul's physical container had been reduced to mush? Why did maggots feast on the flesh of the sinful and the virtuous alike? These anxieties gave rise to a number of poems that play out the encounter between the human body and its vermicular assailant as a form of debate. In the Old English poem "Soul and Body" (*c.* 900 CE), found in the Vercelli Book, the soul addresses its body after death, condemning it for its past sinful actions. The poem offers a parallel image of the carnal appetites that the body indulged in while it was alive in the gorging of a worm, which is given the name "Gifer" [glutton]:

Hwæt, ðu huru wyrma gyfl
lyt geþohtest, þa ðu lustgryrum eallum
ful geeodest, hu ðu on eorðan scealt
wyrmum to wiste!

[So, food for worms,
you certainly didn't think much, while you were following all your terrible pleasures,
about how you will have to be in the earth
a banquet for the worms!]

("Soul and Body I")

In a dynamic that would later be echoed by Shakespeare in Hamlet's cryptic explanation of the dead Polonius's whereabouts—"not where he eats, but where 'a is eaten"—the poem suggests that the body is now consumed by its own base appetites (Blakemore Evans 1997: 1219).[4] The grotesque parallel that "Soul and Body" evokes between base desires and wormy greed is reflected in the repetition of the word "wiste," which is used to describe both human and vermicular feasting.

Operating within the *memento mori* tradition, the poem's macabre descriptions of physical decay appear designed to shock Anglo-Saxon audiences out of their sinful behaviors, encouraging them to repent for their sins before it is too late. The conversation between body and soul continues in Middle English dialogues such as "The Desputisoun bitwen þe Bodi and þe Soule" [The Dispute Between the Body and the Soul], also known as "Als I Lay in a Winteris

Night" [As I Lay on a Winter's Night]. This popular poem, which appeared in a number of miscellanies from the late thirteenth century onwards, consists of a debate between the body and the soul about which of them was responsible for the sins committed during the pair's lifetime (Linow 1970; Rooney 2007: 160–2). A further twist on this genre is "A Disputacioun betwyx þe Body and Wormes" [A Dispute Between the Body and Worms] (*c.* 1460–1470), hereafter referred to as the "Disputation." The poem is told from the perspective of a pilgrim who falls asleep in a churchyard, and in his slumber overhears a debate between the corpse of a beautiful woman and her vermin. The woman laments her posthumous state of decomposition, asking why after leading a noble and pious existence, she has been dealt the same fate as criminals and sinners.

The worms explain that all men and women receive the same treatment, and that throughout her life she has already been beset by "lyce or neytes in þi hede alway | Wormes in þe handes flees in þe bedde" [lice or nits in the head always | worms in the hands, flees in the bed], which were "mesyngers" [messengers] designed to "warne" [warn] the lady of her posthumous fate (Rytting 2000: 223). Rather than preying on the corpse, the worms take on a purgative role, cleansing the body of its rotting flesh:

> we will not lit departe þe fro
> To we hafe scowred and pollysched to
> And made als clene as we can
>
> [we will not depart from you
> until we have scoured and polished
> and made you as clean as we can]
>
> (Rytting 2000: 221)

Despite the emphasis on "orrybyll flesche rotyng and stynkynge" [horrible flesh rotting and stinking], the worms are instrumental to the removal of the defects and diseases of the human soul, enabling the woman to strip away some of her illusions about herself as she realizes that she has suffered from vanity and pride over the course of her life. The poem transforms the vengeful worms of earlier debate poems into a force of redemption that helps the human subject to recognize and atone for her sins retrospectively. One reason for this shift in cultural attitudes towards worms is suggested at the beginning of the poem, when the speaker describes embarking on a pilgrimage away from the "huge mortalite | Of sondre disseses with þe pestilence" [vast loss of life | due to diseases associated with plague]. In a period of unprecedented death and decay, worms appear to have lost some of their previous potency in the medieval imagination as symbols of moral degradation.

The transition from the Old English "Body and Soul," in which the worm mimics the diabolical appetites of its human host, to the late medieval "Disputation," in which the worms purge the lady of her vices, is suggestive of a development in cultural attitudes towards these lowly life forms. Indeed, the substitution of "Soul" for "Worms" in the poem's title may be linked to the gradual overturning of the *contemptus mundi* tradition from the late fourteenth century onwards. As Georges Duby has observed, a growing acceptance of the materiality of the flesh in late medieval writing is reflected in the more realistic depictions of the body in art and sculpture (Duby 1993: x–xi). By the end of the poem, the woman has not simply accepted her carrion, but she even embraces them, concluding: "Lat vs be frendes ... Lat vs kys and dwell to gedyr euer more" [Let us be friends ... Let us kiss and dwell together forever] (Rytting 2000: 220, 225). The speaker's attitude is further evidenced by the marginal illustrations that accompany the manuscript, in which the corpse holds out her hand to a group of worms hovering above her in a gesture of religious reverence (Figure 7.2).[5]

The role played by the worms of "Disputation" in not only destroying but also in reforming previous conceptions of the materiality of the human body is reflected in their ability to unmake and remake the textual corpus. An Anglo-Saxon riddle in the Exeter Book reads:

Moððe word fræt. Me þæt þuhte
wrætlicu wyrd, þa ic þæt wundor gefrægn,
þæt se wyrm forswealg wera gied sumes,
þeof in þystro, þrymfæstne cwide
ond þæs strangan staþol. Stælgiest ne wæs
wihte þy gleawra, þe he þam wordum swealg

[A moth ate words. To me that seemed
a fantastical event, when I found that wonder out,
that a worm swallowed certain of men's word,
a thief in darkness, a glorious statement
and strong its foundation. The thieving stranger was not
a whit more wise that he swallowed those words]

(Krapp and Dobbie 1936: 205)

As Martin Foys notes, most critics agree that the solution to the riddle is some version of "book-worm," or "book-moth," "a page-eating insect we today would recognize as a larval form of the death watch beetle (*Xestobium rufovillosum*), ... the common furniture beetle (*Anobium punctatum*)," or the "booklouse (*Trogium pulsatorium*)" (2018: 2). Yet by omitting all reference

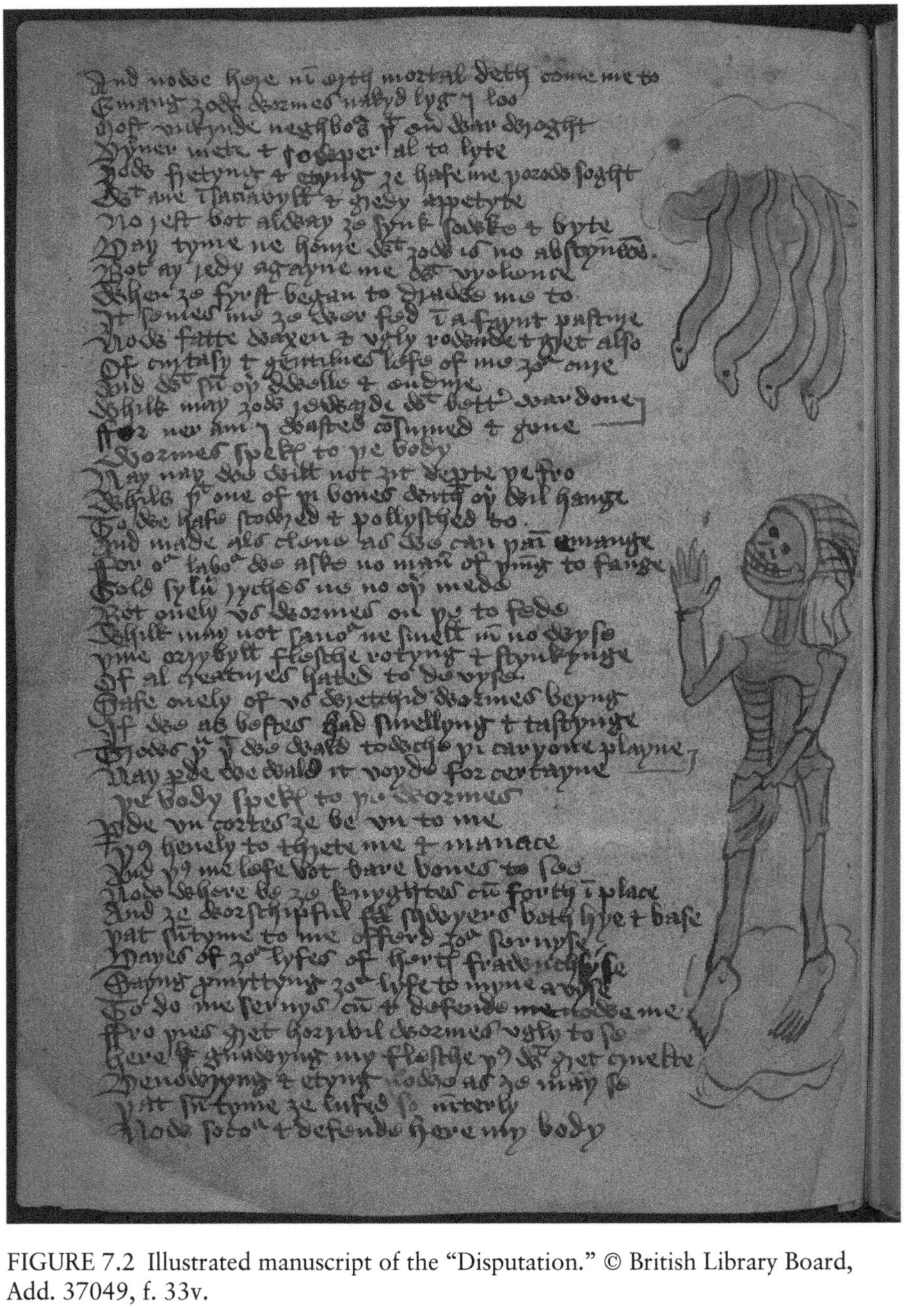

FIGURE 7.2 Illustrated manuscript of the "Disputation." © British Library Board, Add. 37049, f. 33v.

to the book as the object that the "Moððe" consumes, the riddle also hinges on the ambiguity of the noun "word," which may refer to either written or spoken language. In order to make sense of the perplexing image of a moth eating the words out of a man's mouth, we are not able to simply "swealg" [swallow] the words that we are given; rather, as readers it is necessary to chew things over, ruminating over the semantic possibilities of what is being presented in deceptively straightforward language. The poem invites us to resolve the meaning of something that contains holes of various kinds—gaping mouths, moth-eaten sections of missing parchment, the missing word at the heart of the riddle—and yet in our squirming efforts to make sense of this worm-eaten assemblage, a host of new semantic possibilities are generated.

One way of understanding the text, for instance, is that it is based on a common understanding of insects as creators as well as the destroyers of the written word. A pigment that was often used in the creation of medieval manuscripts was iron gall, the principal ingredient of which was oak gall—a curious ball-shaped lump that grows on the leaves and twigs of oak trees. It is formed when a gall wasp lays its egg in the growing bud of the tree, and a soft, apple-like sphere begins to form around the larva. When crushed and fermented, oak gall created the basis of the deep black drawing ink used throughout the Middle Ages (Clemens and Graham 2007: 19–20). The red pigments that were used in the headers and titles of medieval manuscripts also owe a debt to insects: the scarlet ink "cochineal" is made from the ground shells of *Dactylopius coccus* scale insects, while the crimson shade "vermilion" derives its name from the Latin "vermes," or worm, possibly due to its similarity to the deep red color produced by cochineal insects (Eastaugh et al. 2004: 210–11). Similarly, the crimson dye known as "kermes," which takes its name from the beetle from which it is obtained, was used in the medieval era for dying silk (also produced by worms) and wool. Kermes was a popular shade for medieval bindings (Shailor 2002: 63; Glick et al. 2005: 380), which suggests that insects were responsible for many of the materials involved in the production of medieval texts.[6]

Regardless of their creative powers, worms retained primarily negative associations in the Middle Ages: in *The Canterbury Tales* (1387), Geoffrey Chaucer's account of the fate of Jewish persecutor King Antiochus IV—"The wrecke of God him smote so cruelly | That through his body wikked wormes crepte" [He was punished so cruelly by the wrath of God | That wicked worms crept through his body] (Benson 2008: 250)—indicates that these life forms remained bound up with notions of suffering and sin. The persistence of the worm as humankind's internal destroyer is also suggested by the Middle English figure of speech for madness—"to have a wild worm in one's head"—which was used long into the early modern period (Whiting and Whiting 1968: 675). As the next section will suggest, the lingering malignity of these creeping things may be contrasted with a creature that maintains a far more elevated status in the medieval imagination.

APIAN ALLEGORIES

No insect was prized more in the Middle Ages than the western honey bee, or *Apis mellifera*. The apparent selflessness, relentless industry, social cooperation, and seemingly sexless reproduction of bees, coupled with the sweetness and purity of their products—honey, mead, wax—led many to discern parallels between these winged creatures and members of the Christian Church (Ransome 2004: 144–9; Preston 2006: 77–81). Odo of Cheriton's fable "The Bees and The Beetles" likens the former to "doctors of the church, contemplatives, who invite in an impious man and set before them combs of honey – i.e. the Lord's precepts as well as His law" (Jacobs 1985: 103). Their apparent chastity also caused some writers to draw comparisons with the Virgin Mary: the tenth-century lyric "The Voice of the Turtle" concludes: "Yet not the birds, but bees afford | The pattern of sweet chastity | Like her who bearing our Christ Lord | was undefiled" (Jones 2013: 258).

Medieval preconceptions regarding the virginal status of bees largely derived from Latin translations of Aristotle's *History of Animals* (350 BCE), which alleges that these creatures do not copulate like other animals, but instead fetch their young from flowers. Aristotle also noted that there are worker-bees and drones whose sole job it is to protect the king (i.e., the queen-bee), that bees do not tend to attack except if their hive is disturbed, and that the hive is kept meticulously clean (Balme and Peck 2014: 187–201). Aristotle's account of the many virtues of bees influenced a number of early Christian thinkers, including the saints Ambrose, Augustine, Jerome, and Basil, as well as the thirteenth-century theologians Thomas of Cantimpré and St. Antony of Padua, the latter of whom referred to the risen Christ as an *apis aetherea*, or celestial bee (Ransome 2004: 145–6). In his homily "The Bee and the Stork," the fourteenth-century mystic Richard Rolle argues that:

> Thay kepe thaire wynges clene ... and thay hafe othyre vertus vnblendyde with þe fylthe of syne and vnclene luste. Arestotill sais þat þe bees are feghtande agaynes hym þat will drawe þaire hony fra thayme. Swa sulde we do agaynes deuells þat afforces thame to reue fra vs þe hony of poure l[u]fe & of grace.
>
> [They keep their wings clean, and they have other virtues unblended with the filth of sin and unclear lust. Aristotle says that bees only fight against those that try to take their honey from them. So should we fight against devils that try to forcefully take away from us the honey of pure life and grace.]
>
> (Richard Rolle, "The Bee and the Stork," in Horstman 1851)

Like the fable, the passage moves from an account of insect behavior to an *epymithium*, or moral lesson, in which these actions are translated into the realm of human conduct. Throughout the sermon, Rolle presents the hive as a model for Christian society, in which its workers are "neuer ydyll" [never idle], its most vulnerable members are never neglected, and all are united in love of and subservience to a benign leader, "Ihesu Criste" [Jesus Christ] ("The Bee and the Stork").

Bees also function as allegories of Christian virtue: in the fifteenth-century biblical narrative "The Storie of Asneth" (*c.* 1450–1475), the "bred of lyf" [bread of life] resembles a piece of honeycomb from which bees of Paradise emerge "white as the snow; | Here wynges were of purpre hewe" [white as snow; | their wings were of a purple hue] (Peck 1991: li. 552, 567–8). Here, the sacramental act of converting one substance into another—flowers to honeycomb, bread to body—renders the bee a fitting emblem for transubstantiation. The early fifteenth-century poem *Mum and the Sothsegger* (1403–1409) also features an extended dream vision in which the "Sothsegger" [truth-seeker] King is depicted as a beekeeper presiding over unproductive drones that steal honey from other hardworking bees:

> For whenne the bee-is bisynes is bribed fro the hyve
> Thorough dranes that deceipuen thaym and doon no thing elles,
> Thenne seen the bees thair subtilité and serven thaym thereafter
> As Bartholomew the Bestiary bablith on his bokes

FIGURE 7.3 Detail of a miniature of bees collecting nectar and returning to their hive, from a bestiary with theological texts, England, *c.* 1200–1210. © British Library Board, Royal MS 12 C XIX, f. 45v.

[For when the busy bee is lured away from the hive
By drones that do nothing more than deceive them,
Then continue to lie to the bees by working for them afterwards
As Bartholomaeus Angelicus tells us in his books]

(Dean 2000: li. 1051–4)

The alliterative verse of the poem echoes the productive buzz of the beehive, as is conveyed by the lines such as "The bees been so bisi," as well as the euphonic hum of "And helpen to make hony of that thay home bringen." Form and content are thus harmonized in this passage, which not only mimics its subject matter, but also converts the ineffable buzzes and hums of apian activity into intelligible signs. Here we might recall Alain de Lille's exegetical account of the animal kingdom as "like a book and a picture | to us" that is able be read and interpreted by human beings in terms of its moral and spiritual symbolism.

"The Storie of Asneth" diverges from early scientific accounts of bee anatomy and behavior by naturalists such as Aristotle, Pliny the Elder, or Isidore of Seville: the white bees with purple wings that generate from honeycomb are dressed in symbolic Christian garments—white for purity, purple for piety—and the passage is most likely an allusion not to natural history but to a passage in the New Testament in which the resurrected Christ eats a piece of honeycomb (Luke 24:42). In *Mum and the Sothsegger*, the source provided is Bartholomaeus Angelicus's medieval bestiary *De proprietatibus rerum* [*On the Properties of Things*] (1240), which is heavily indebted to Aristotle's observations of bees, and yet the behavior of bees in this passage exists in stark contrast to Bartholomaeus's insistence that "none of [the bees] aspyeþ ne waitiþ to taken out of othire's trauaile, noþir takeþ wrongfulliche, noþir steliþ mete" [none of them spy on each other or take credit for another's work, take wrongfully, or steal from each other] (Seymour 1975: 611). Clearly then, "The Storie of Asneth" pays lip service to natural history texts in order to lend weight to its interpretation, but like the majority of medieval texts it is more interested in exploring the allegorical function of bees than in remaining true to empirically grounded studies of their behavior.

The allegorical function of bees in medieval literature is also evident in Dante Alighieri's *Divine Comedy* (1320), which traces the poet's journey through the various layers of the underworld. Although insects of all descriptions appear in the *Inferno* and *Purgatorio* sections of the text, only bees grace the pages of the *Paradiso*. The *Inferno* describes the "swarms of wasps and hornets" who sting the bodies of the dead, while the *Purgatoria* likens the undeveloped souls of those in purgatory to "defective insects | Like the grub which has not reached its full development" (Sisson 2008: 57, 242). In the *Paradiso*, on the other hand, Dante likens the "holy army" of angels to "A swarm of bees, landing upon flowers | At one moment and then at once returning | To where its work is turned into sweetness" (Sisson 2008: 486). Here, the labor of bees is figured as an act of divine transformation, as the "work" involved in "sing[ing] | The glory of him who fills them with his love" is converted into sweet rewards.

Dante is participating in an epic tradition that extends back to Homer's *Iliad*, in which the Achaean armies are likened to "swarms of clustering bees that issue forever | in fresh bursts from the hollow stone, and hang like | grapes as they hover beneath flowers in springtime" (Lattimore 1976: 78). Functioning as a form of superorganism, the army take on the appearance of social insects acting in accordance with a collective rather than an individual will. Dante's bee simile also harks back to the writing of the Roman poet Virgil, who appears as the author's guide through the underworld. In Book 6 of his epic poem *The Aeneid* (19 BCE), Aeneas reaches a sequestered grove in the underworld, in which "hovered peoples and tribes unnumbered; even as when, in the meadows, in cloudless summertime, bees light on many-hued blossoms and stream round lustrous lilies and all the fields murmur with the humming" (Fairclough 1994: 554).[7] Aeneas's father Anchises informs him that these apian spirits are waiting to be reborn as the future leaders of Rome, and, likewise, in Dante's version there is a similar sense that bees represent a kind of spiritual self-renewal, and a return to earth, as is reflected in the parallel imagery of "light[ing] on ... blossoms" and "landing upon flowers."

Dante's bee simile may also have been inspired by Virgil's agricultural poem *The Georgics* (29 BCE), which contains a chapter devoted solely to beekeeping. As well as presenting the hive as an ideal model for a human society based on loyalty, selflessness, and productivity, Virgil outlines a miraculous phenomenon that influenced many medieval writers and thinkers, known as *bugonia* (Fairclough 1994: 216–18). Deriving from the ancient Mediterranean belief that bees are able to generate spontaneously out of a dead bull's carcass, the concept is also present in Samson's Old Testament riddle: "Out of the eater, something to eat; out of the strong, something sweet" (Judges 14:14). Together with classical accounts of the chastity of bees, this account of apian rebirth demonstrates the ways in which the exalted status of bees in medieval culture relied on the circulation of beliefs that, although subsequently disproved by science, imbue these creatures with otherworldly talents. Just as in classical mythology, birds were often perceived as winged messengers from on high, the quasi-mystical connotations of bees as aery sky-dwellers, combined with their skills as miraculous procreators and figures of spiritual renewal and rebirth, elevated them to dizzy heights in the medieval imagination.

Not all writers, however, held bees in such high esteem. In his prologue to "The Summoner's Tale," Chaucer delivers the following account of friars:

> Right so as bees out swarmen from a hyve
> Out of the develes ers ther gonne dryve

Twenty thousand freres on a route
And thurghout helle swarmed al aboute,
And comen agayn as faste as they may gon,
And in his ers they crepten everychon.

[Much as if bees were swarming from a hive
Some twenty thousand friars in a rout
And swarmed all over Hell and round about
And then came back as fast as they could run
And crept back into [Satan's] arse again, each one.]

(Benson 2008: 128)

As we shall see, Chaucer was not alone in likening friars to insects, and here the image of the swarm is used to evoke a model of social disorder, rather than of collective harmony, in which bees are destined to congregate in "Hell" as diabolical entities rather than divine beings. Chaucer's simile evokes the more negative connotations of the verb "swarm," which can be traced back to the Norwegian dialect word "svarva," which describes agitated or confused movement, as well as the German "schwärmen," meaning to riot or rave (*Oxford English Dictionary* 2018). It is possible to hear the hum of Dante's paradisal bees, as well as to the apian allegories of medieval sermons. Chaucer's use of epic simile therefore appears designed to subvert and undermine the image of the friar as a pious figure at an intertextual as well as a figurative level by debasing a creature commonly associated with Christian virtues.

As well as being the first known writer to use the word "swarm," Chaucer's work also contains the first recorded instance of the figure of speech "as busy as a bee," when the Host of his *Canterbury Tales* describes women as "bisy as bees | Been they, us sely men to deceive" [they're as busy as bees | to deceive us silly men] (Benson 2008: 168). The connection forged by the text between apian activity and human dissemblance is also suggested in "The Second Nun's Tale," when Chaucer describes the pious Cecilia as "lyk a bisy bee" [like a busy bee] before adding: "withouten gyle" [without guile] (Benson 2008: 265). Chaucer's preoccupation with apian craftiness may be related to long-standing cultural associations between bees, honey, and the craft of speech. In Hebrew the word for "bee," *dbure*, derives from the root *dbr*, meaning "speech," and Sophocles, Plato, Pindar, Virgil, and Lucan are all said to have "had their lips touched with honey" during infancy, before going on to become skilled rhetoricians and poets (Ransome 2004: 103–5; Winston 2014: 140). St. Bernard of Clairvaux was also widely referred to as a "mellifluous doctor" (from the Latin "mel," honey) due to his gift for language, and a legend surrounds St. Ambrose stating that a bee settled on his lip as a child, thus marking out his future eloquence (Preston 2006: 13–15).

The parallels between apian productivity and the literary hive mind were also emphasized by the Italian poet Petrarch, who remarked:

> It may all be summed up by saying with Seneca, and with Flaccus before him, that we must write just as the bees make honey, not keeping the flowers but turning them into a sweetness of our own, blending many very different flavours into one, which shall be unlike them all, and better.
>
> (Robinson 1969: 291)

Petrarch is suggesting that the poet borrows from existing material like a bee flitting from flower to flower, gathering up various literary sources before distilling them into fragrant new blends. This analogy also contains an allusive murmur of Pindar's account of the poet as a bee culling his songs from the Muse's garden to produce poetry "sweeter than bee-built honeycombs" (Race 1997: 395; Liebert 2010: 98). Similarly, a Latin gloss to John Gower's Middle English poem *Confessio Amantis* [*The Lover's Confession*] (1390), which may also have been written by the author himself, likens the poet to a worker bee gathering nectar, learning the craft of writing through the cross-fertilization of classical influences (Gastle 2010: 184–5).[8] These analogies of honeyed eloquence were also tinged with negative associations: in *The Republic*, Plato associates honey with the dangers of poetic language as a source of pleasure that is inimical to reason and order in society, warning that "if you grant admission to the honeyed muse in lyric or epic, pleasure and pain will be lords of your city instead of law and that which shall from time to time have approved itself to the general reason as the best" (Shorey 1969: 607a).

A passage in Proverbs 5:3 also associates this sweet nectar with the corruptive influence of pleasure, describing how "the lips of the adulterous woman drip honey, and her speech is smoother than oil; but in the end she is bitter as gall, sharp as a double-edged sword." The honey-tongued speaker is thus a double-edged figure, representative both of exemplary rhetoric and the flagrant misuse of language. This notion persists into the Middle Ages: in his cautionary "Tale of Melibee," which features an apian-named protagonist who gorges on earthly pleasures, Chaucer quotes Ovid's warning that: "under the hony of the godes of the body is hid the venim that sleeth the soule" [under the honey of the gods of the body is hidden the venom that kills the soul] (Benson 2008: 229). Honey then, is associated with the dangers of physical appetites, as well as the risk of succumbing to sweet-sounding words. Despite the predominantly pleasing associations forged between bees and ideas of purity and eloquence, there is also a sting in the tale.

FLEAS, FLIES, AND FRIARS

If bees were among the most esteemed of all animals in medieval society, then flies and their allied ephemera—fleas, gnats, lice, even butterflies—were positioned at the opposite end of the scale, often signifying the total absence of value. In Chaucer's "The Shipman's Tale" a discontented wife asserts of her husband: "he is nought worth at al | in no degree the value of a flye" [he's not worth the value of a fly], while in "The Reeve's Tale," the mischievous Alan asserts of the miller: "Him! I coont him nat a flea!" [Him! I value him less than a flea] (Benson 2008: 205, 82). In the latter phrase in particular it is possible to hear not one but two worthless critters—both "nat" [not] and the Middle English "gnatte" [gnat]—a pun that echoes the way in which Chaucer's writing utilizes these lowly creatures to confer negative value upon individuals. This process of devaluation also applies to the words of others: the "boring" monk's tale is deemed "nat worth a boterfly" [not worth a butterfly] by the assembled pilgrims, while the wife of Bath asserts that she will pay as much attention to men's instructions about how women should behave "as a gnat" (Benson 2008: 252, 109).

The diminutive size, frailty, ephemerality, and lack of obvious use value of flies, fleas, and gnats is consistent with their association throughout human history with qualities of insignificance. Flies in particular seem to scarcely warrant attention, remaining instead a mere background buzz, or even a category of inert matter. Isidore of Seville reinforces this connection by discerning a false etymological link between fleas and dust: "fleas (*pulex*) are so called because they are mainly nourished by dust (*pulvere*)" (Brumble 1973: 148; Barney et al. 2006: 259). This resemblance is also suggested by the words "mite" and "mote," which the *OED* suggests may derive from the shared Germanic base "moth," with the original sense "to gnaw, reduce to small fragments" (*Oxford English Dictionary* 2018). The connection between insects and pulverized particles is also implied by the etymology of the Greek word "atom," meaning "indivisible, not capable of being divided," and "entomos," or "cut into sections," from which Aristotle coined the Greek word "entoma" [insect] (Connor 2006: 90).[9]

Another entomological connection during the Middle Ages that appears to partly hinge on an linguistic resemblance is that of fleas, flies, and friars, or rather "Flen, Flys, and Freris" (*c.* 1475)—the title of an anonymous invective against the Carmelite brothers of Cambridge. The poem echoes the imagery of friars as a Satanic swarm in "The Summoner's Tale," as well as Chaucer's subsequent analogy of these figures as greedy diptera: "A frere wol entremette hym everemo. | Lo, goode men, a flye and eek a frere | Wol falle in every dyssh and eek mateere" [These meddling friars are always on the prod! | Don't we all know that a friar and a fly | Go prod and buzz in every dish and fly] (Benson 2008: 116). In a macaronic combination of Latin, English, and coded language, "Flen, Flys, and Freris" continues this anti-fraternal tradition:

Flen, fly3es, and freris, foul falle hem thys fyften 3eris,
For non that her ys lovit flen, fly3es, ne freris.
Fratres Carmeli navigant in a bothe apud Eli,
Non sunt in cœli quia gxddbov xxkxzt pg ifmk.

> [Fleas, flies, and friars – foul fall them these fifteen years!
> For none that is here loveth fleas nor flies nor friars.
> The Carmelite brothers sail in a boat called Ely
> They are not heavenly because they fuck the wives of Ely.]
>
> (Wright and Halliwell 1841: 91)

Implicit in this passage is a long-standing association between flies and the devil: a ninth-century glossary defines Beelzebub as "vir muscurum," or man of the flies (Taylor 2013: 169). The harsh fricatives of the poem also evoke the friction between friars and locals during this period; many felt that the brothers had grown too numerous and powerful by the Late Medieval Period, with the rapid growth of this community resulting in widespread allegations of lechery, heresy, and corruption (Szittya 1986: 3–10; Rayborn 2014: 117–18).

The poem teems with entomological imagery: "Pediculus cum sex pedibus me mordet ubique" [six-footed lice bite me all over], which evokes the invasion of the friars into the body politic, not to mention their agitated, flea-bitten lust. Indeed, associations have often been forged between fleas and sexual desire: the late medieval erotic poem "Elegia De Pulice" ["Elegy of the Flea"], which was initially attributed to Ovid, is addressed to a flea that the male speaker accuses of trespassing into the private regions of a young woman's body:

> when the maid
> lies reclining, you ravage the thigh, and cause the gore
> to flow. And meanwhile you have dared to broach even the
> passionate parts, and to taste the pleasures born in
> those places.
>
> (Brumble 1973: 148)

The speaker's voyeuristic tone, which would later be echoed in John Donne's erotic poem "The Flea," hinges on the intimate proximity between the human host and its puny parasite; the ability of these life forms to explore "even the | passionate parts" of the body, coupled with their similarity of fleas to pubic crab-lice grants them a licentious (pun intended) quality. The poem also captures a key aspect of the medieval fascination with insects, with the male speaker acknowledging his desire "to be immediately transformed," and to "be changeable," moving back and forth between man and flea in order to woo his mistress (Brumble 1973: 148). Just as in Donne's metaphysical poem, the flea functions as a conceit for the forbidden pleasures of sexual intercourse—"in this flea, our two bloods mingled be"—the speaker conceals the full extent of his lust beneath the guise of a diminutive transgressor (Smith 1996: 58).

Chaucer's dream vision, *The Legend of a Good Woman* (*c.* 1380s), celebrates the noble figure that can rise above such trivial nuisances:

For loo, the gentil kynde of the lioun:
For whan a flye offendeth him or biteth,
He with his tayl awey the flye smyteth
Al esely, for, of his genterye,
Hym deyneth not to wreke him on a flye,
As dooth a curre or elles another beste.

[Consider for instance the gentle nature of the lion
When a fly provokes or bites him
He simply swats it away with his tail
Effortlessly, for his nobility means that
He has no desire to kill that fly
As a villain or some other beast might.]

(Benson 2008: 599)

As well as suggesting that noble individuals should not avenge themselves on lesser creatures, the moral of this passage seems to be that it is not worth wasting your energy on something so insignificant and relatively harmless—a mere fleabite. This analogy demonstrates that unlike worms and other verminous entities, fleas and flies remain superficial creatures in the medieval imagination that exist on the surface of the body rather than tangled up in the innards, functioning as minor nuisances rather than deadly menaces. Indeed, it was not until the nineteenth century, when the spread of the bubonic plague during the Middle Ages was finally attributed to the fleas that lived on the bodies of rats, that people realized that these creatures could be deadly (Gross 1995: 7609–11; Byrne 2012: 328–9, 366–7).

A few writers did seek to foreground the more malign qualities of flies: in his "Vituperio Della Mosca" [Condemnation of the Fly], the twelfth-century translator and poet Eugenius of Palermo launches a sustained attack against this "animaletto così miserabile e sudicio" [wretched and dreadful little animal] (Gigante 1964: 152).[10] The poem characterizes the fly as a cowardly creature, a disruptive nuisance, and the most gluttonous of all creatures. This "mostro vergognoso di meschina formazione, di breve vita" [shameful beast of meagre form and short-lived existence], the poet argues, doesn't even deserve to be classified as an animal, but rather simply part of the formless chaos of the underworld (Gigante 1964: 152). Flies also earned the respect of some medieval writers: Odo of Cheriton tells the story of a religious sceptic who exhorts that God couldn't have made a creature so foul. This heretic is then bitten in the face by a passing fly, and the fable praises the creature for avenging the slight delivered to his divine creator (Jacobs 1985: 82). Humanist writer Leon Battisti Alberti's Latin playful encomium "La Mosca" [The Fly] (*c.* 1441–1443) also pays tribute to the strength, persistence, acuity, and audacity of flies, drawing on Homer's account in *The Iliad* of the daring of a fly that "though it is driven hard away from a man's skin, even | so … persists in biting him" (Lattimore 1976: 369; Connor 2006: 20–4).

Eugenius' and Alberti's poems were inspired by the Roman satirist Lucian of Samosata's "Praise of the Fly" (1 CE), which deploys the rhetorical trope of hyperbole to elevate this lowly creature to the level of a much larger animal. The poet even playfully acknowledges this strategy when he concludes: "I will stop talking … for fear you may think that, as the old saying goes, I am making an elephant out of a fly" (Hendersen 1913, "The Fly": 95). The text is fly-like in its movements, alighting rapidly on a series of Classical allusions to diptera, and whizzing from one argument to the next in a manner that seems designed to disorient readers, and ultimately to pester them into submission. Alberti's poem attempts a similar reversal of attitudes by arguing that flies are much more impressive than bees, who store up their food rather than feasting together, and who are presided over by a tyrannous king (i.e., the queen bee). The humble fly is not selective about the company it keeps: "infatti abbraccia e bacia allo stesso modo il principe e il plebeo, il ricco e il povero, sfiorandoli con le ali e applaudendoli" [it embraces and kisses the prince and the plebeian, the rich and the poor alike, touching them with its wings and applauding them] (Contarino 1984: 181).[11] The text challenges dominant conceptions of flies as lowly pests, encouraging readers not to simply swat away these pesky creatures, and instead to allow themselves to be touched by their unexpected talents.

Alberti's affectionate anthropomorphism is also demonstrated by his argument that flies witness all manner of human sins with their enormous eyes, and yet they remain paragons of discretion. In a further attempt to recuperate the much-maligned status of the fly, he notes:

> O pigri e lenti, che a stento anche con il pungolo la mosca riesce a spingere all'azione, imparate i buoni costumi dalla solerte mosca, maestra di virtù! La mosca non sta mai in ozio e stimola, come può, i pigri al loro compito.
>
> (Contarino 1984: 191)

> [Oh you lazy and slow, with hardly a push the fly springs to action, learn the good habits of the diligent fly, teacher of virtue! The fly never dwells in idleness and pushes, as it can, the lazy to their work.]

The fly is thus not so much a nuisance as a stimulant—an enlivening presence in human affairs. Alberti's pronouncement of the virtues of the "solerte mosca" [diligent fly] is clearly tongue-in-cheek, with the poem generating a humorous disparity between its high-flown style and lowly subject matter. Yet it also demonstrates that, centuries before the invention of the microscope, writers were looking closely at these domestic inhabitants, and observing their multi-faceted eyes, as well as the "straordinaria figura geometrica" [extraordinary geometric figure] on their wings, "la misura dell'altezza di tutti i poligoni" [the measure of the greatness of all polygons] (Contarino 1984: 189). Such detailed attentiveness to the insect world is also evident in Dante's *Divine Comedy*: the movement of a dragon is likened to a wasp "which pulls back its sting | Drawing its magnificent tail to himself," and when a crowd jostles closely in the Underworld, the poet remarks: "just as within their dark companies | Ants rub noses with one another" (Sisson 2008: 312, 342). Although these similes may not be anatomically correct—ants, after all, do not have noses—they evince a remarkably intimate entomological knowledge.

The insects of medieval literature are complex and multifaceted creatures, freighted with moral and religious symbolism, bound up with Classical allusion, and inextricable from the at-times far-fetched observations of early naturalists. Yet what unites these entomological examples is a fascination with the changeability of insect life: as we have seen, it is the capacity of these creatures to radically alter and reshape the human body, to disrupt settled hierarchies, and to expand the possibilities of language itself by stimulating new forms of wordplay, figures of speech, and metaphors that explains their particular appeal to medieval writers. In examining the presence of insects in the writing of this period—from Anglo-Saxon riddles to late medieval lyrics, from Old English words to Middle English proverbs—it becomes clear that these creatures have burrowed deep into the medieval imagination and have become embedded, literally as well as figuratively, in the fabric of our language.

CHAPTER EIGHT

Insects in Art

JULIO FERRER

In artistic terms, the historical era called the Middle Ages refers to the period from the fall of Rome (476) until the discovery of America (1492). The Middle Ages began with the decay of the Roman Empire, the systematic destruction of the ancient artistic heritage, and loss of knowledge of the anatomy and the laws of perspective in art. On the contrary, the Renaissance—which immediately followed the Middle Ages—was a period of artistic research and discovery: the theory of proportions, the central perspective and geometric compositions were developed, and appeared, clearly formulated, in diverse books by artists such as Leonardo de Vinci and Albert Dürer, coinciding with the first anatomical studies performed by Andrea Vesalius. This artistic period culminated at the end of the sixteenth century with the eruption of a new style, the *Baroque*, which combined previous Renaissance influences with new studies on the effects of sunlight and shadows and the illusion of movement, and introduced radical changes in composition and new techniques, such as combining linseed oil with egg emulsion to create a new painting material that intensified the visual impact of the paintings.

In art, a parsimonious method of study is the iconographic analysis introduced by Panofsky (1939) on how to deal with problems related to art manifestations in mythology and folklore (Ferrer 2009a, 2014). This analytic method is dissociative, focusing separately on every iconographic element in order to establish its historical sources, symbolic use, and significance as a whole as related to the artist. When different elements of the work are studied and compared independently, often a plausible interpretation of the historical sense and the intentions of the artist emerges. In approaching the theme of entomology in medieval art, not only insects but also some other arthropods, worms, and fantastical creatures must be taken into account, since during the Middle Ages, the modern notion of insect (as laid out by Linnaeus' classification in *Systema Naturae*, 1758) did not exist. Even Linnaeus himself considered centipedes, arachnids, and scorpions as insects.

THE SOURCES OF RESEARCH

Today many illustrated manuscripts that were produced during the Middle Ages (fifth–fifteenth century CE) are readily available online. The margins of these manuscripts are often ornamented with accurate decorative illustrations, representing plants and animals, also known as "marginalia." However, many legends, ideas, and superstitions related to insects are so old that the sources are unknown and literary references do not exist. In Sweden, oral legends have been compiled since the end of the eighteenth century in a national institution, the Swedish Archive (*Folkminnessamling*). This curious archive, inspired by the Brothers Grimm's research of popular tales, contains manuscript cards with notes written by ethnographs and folklorists since the end of eighteenth century, reporting orally transmitted histories and anecdotes from generation to generation in various regions of Sweden, Norway, and Finland. The archive is ordered thematically; for example, searching "insect" or "beetle," etc., reveals several testimonies, accurately annotated with themes, time, and locality, and the names of the informant and the collector of information. This archive, which includes not only oral traditions but also notes from journals, pamphlets, posters, etc., is a remarkable source of information.

BESTIARIES AND ILLUMINATED BOOKS

Many insects were used as ornaments to decorate pages or as illustrations in medieval books and manuscripts, but compared to other animals, illustrations of insects are relatively uncommon. As already stated, the modern concept of insect was not yet established and for a long time many arthropods were treated as kinds of "worms." The iconography of insects and other arthropods during the Middle Ages is mostly unrealistic, representing many symbols taken from different sources. Many insects appear, with different significance, in the Bible or other religious texts, in literature, or in the ancient fables.

Generally, with the exception of butterflies, bees and ants, most insects were considered hideous, repudiated as harmful and associated with evil. They were treated as highly repulsive, created exclusively to punish the enemies of the God of the Old Testament. For this reason, compared with the situation in the following centuries, the interest and knowledge of insects during the Middle age is not only poor, but often bizarre.

One of the best sources for understanding the medieval concepts of insects and other animals are the Bestiaries. Known as the medieval books of zoology (Hersig 1995), these illuminated books presented the medieval interpretations of insects and other animals, compiled and illustrated from different sources. Perhaps the most celebrated bestiary is the one written by the Spanish Bishop Isidore of Seville (560–636). Below is a checklist of insects and related creatures that appear commonly in medieval bestiaries.

Ants. In *Physiologus*, a bestiary from England dated 1236, ants are depicted as giant dog-like creatures on an anthill, digging for gold in Ethiopia (Figure 8.1). These monstrous ants killed anyone who attempted to steal the gold from them, and were the mythical *ant-lion*, described in some ancient sources (Gerhardt 1965). Ants were also considered the proof of the existence of a Divine Providence, and according to the theologian Johannis Nider's work *Formicarius* (The Ant Hill) (Nider 1484) their nest represented the image of an ideal Christian community.

Bees were considered the smallest of birds (Figure 8.2), and according to classical works, they spontaneously generated from oxen carcasses. Bees were among the few insects regarded positively and domesticated. The beehive was an integral part of the human world: in some Italian communities, bees even participated in the familial mournings, and the beehive was covered with a black cloth (Di Nola 2006). Rarely, bees were used to torture Christians martyrs.

Butterflies represent a common decorative, ornamental element in illuminated manuscripts produced during the Middle Ages (fifth–fifteenth century CE). They were used by knights as targets for military, archery, and other weapons exercises: Nazari (2014) gives several interesting examples of

FIGURE 8.1 Ants. Gold-digging ants, as dogs attacking presumptive thieves. Queen Mary Psalter, England, thirteenth century. Royal MS 2 B VII, f. p6r. Tracing drawing by the author.

FIGURE 8.2 Detail of a miniature of bees guarding their hive against a bear. Flore de Veru et Costumes, Padua, Italy. Ink drawing by the author.

FIGURE 8.3 A knight threatens to kill a butterfly. Romance of Alexander, France, Flemish. Bodleian Library, England Oxford, MS Douce, 366, f. 38.

this curious combat (Figure 8.3). On the other hand, caterpillars are symbols of metamorphosis, transforming from a larva to a cocoon stage and then to a completely different creature, emerging as moth or butterfly. Curiously, in medieval artwork, the normal wings of birds were an attribute of the angels and archangels, while insects' wings appear in some representations of demons, such as the works by Martin Schönhauer and Hieronymus Bosch.

BEETLES: THE "GOBLIN" CREATURES

In German mythology, some beetles were called "goblin" (*koboltic*) creatures, associated with dwarfs, gnomes, and other guardians of hidden, subterranean treasures, working with noble metals and precious stones. These fantastic beings were able to make fabulous pieces of jewelry. In many myths and legends, compiled by Bächthold Stäubl (1927–1942) in his monumental work *Handwörterbuch des Deutchen Aberglaubens*, the secret workers of the under-earth were collectively referred to as "Käfer," meaning beetles.

Curiously, many legends and concepts associated with beetles or insects in isolated villages of northern Europe present clear connotations and similarities with popular myths occurring in the Basque provinces of southern France and Spain, compiled by folklorists such as Barandiarán (1960). This is an indication of the remote age of these legends, transcending space and time and probably going all the way back to Prehistory (the Neolithic Period).

Much later, during the Middle Ages, the pagan beetles associated with old divinities like Thor and Freja in northern Europe were Christianized. The beetle associated with rain and storm was the *Thordyfil*, probably because crowds of these beetles appear under favorable circumstances after rain in the summer. The name was later changed to *Thordyvel* (Thor's devil). The beetle associated with Freja was the ladybird, one of the most proliferous insects.

According to Scandinavian legends, Freja was seduced by mythical dwarfs, artisan masters of jewelry. The goddess desired and bought a precious golden necklace from the dwarves in exchange for sex (Baeckstedt 1990). This legend is undoubtedly an early variant of Snow-white, adapted later by the Brothers Grimm to convey the moral values of the German bourgeoisie. This erotic saga indicates that mythical dwarfs and beetles were associated with sexuality and wealth in the old societies of northern Europe.

THE TALES OF THE GOLDEN BEETLE

The "golden beetles" in tales from northern Europe may refer to various species of *Cetonia*, *Potosia*, and *Liocola* (Scarabaeoidea). The old names given to these beetles all contained the word (*guldbagge*, *guldbist*, etc.). Sometimes these beetles were found in ant nests and were thought to be protected by ants and for this reason the golden beetle was named the King of the Ants ("*Myrkonung*") in Sweden. Golden beetles observed in trees and flowers were also associated with fecundity and fertility. Dung was a basic commodity for peasants, and dung beetles were seen as indefatigable workers, carrying dung and fertilizing the soil in the farms. The golden beetles in northern Sweden were reputed to be protectors of cattle, and the first one observed each year was captured, killed, and mixed with flour and salt, and fed to the bell-cow (leader of the herd). The animal was thought to be protected this way from wolf attacks or other accidents in the forest.

According to tales preserved ay the Swedish Archive (NMF), the golden beetle was also a symbol of luck: "*Spiritus* was a little magic box, inside was placed a beetle who needed to be fed every day. I can't remember, but it was something the owner chewed first" (NMF 1884); "The *spiritus* was preserved in a little box and when gold or silver were close, the box moved. In this box was preserved a 'golden beetle' with a little moss and mercury; the box moved towards the place where gold was hidden" (NMF 1868). Another reference says, "all people have a coin in the pants to attract more (coins) to the owner, but Old Holta Sven from Högahult has a bug, an insect in a paper-box. He calls it '*spiritus*' and he says that after nine years, it will 'puke' and make him rich" (NMF 1881). Not only beetles but sometimes frogs and toads were transformed into "spiritus." The owners of these critters were sometimes arrested and processed: Ankarloo (1971) cited a famous prosecution against a presumed witch, a "captain" in the Army of Satan, owner of an artifact: a preserved, dried toad.

Interestingly, similar stories occur in the Basque regions of France. Spiritus ("laminak," literally: sucker) were collected in the forest by leaving an empty box or purse on the forest floor. The mystic creatures would come to the purse in swarms, linger, and ask "Cer eguin? Cer eguin?"—"What are we going to do? What are we going to do?" They are quite demanding, and it is sometimes necessary to sacrifice at least a goose to appease them (Aulle and Douteville 1976: 341). Butterflies also execute orders, as indicated by their Basque name "mandatari" (literally: to be commended).

THE TALES OF THE DUNG BEETLE THORDEVIL

With their black upperside and a surprisingly beautiful metallic aeneous underside, the forest dung beetle (*Geotrupes stercorarius*) (Coleoptera: Geotrupidae) were treated with great respect by the Vikings. These beetles were so prestigious that the name used for males was *tordyfil* (the current name today is tordyvel): the name comes from Tor (Thor), the god of storm and lightning. The reason for this naming is perhaps because dung beetles appear after storms and prefer to excavate galleries in the soil after rain. Every year, *Geotrupes* would be scrutinized to see if the year is going to be a prosperous one or not. If the first-observed tordyvel was alive and running around, then it was a good year. If the surface of body was covered with "lice" (mites, Acari), the year must be prosperous. Contrarily, if the first-observed beetle was free of parasites, it was a sign of starvation; and the worst was to be anticipated if it was seriously injured or dead. If the golden beetle brought wealth for the owner, its brother tordyvel (with its skills in carrying dung to the garden) could fertilize the soil, bring a better harvest, and also good weather.

The *spiritus familiaris*, as the beetles in question, was able to suck milk from domestic animals such as cows and goats. The magical procedure consisted of the production of a "puke" or a "bjära" (*biera*). Demonic creatures—puppets made with a stocking or a sock—carried the beetle inside of them. These magical puppets that "sucked" like a bug were able to go by themselves to neighbors' cows or goats and milk them. The precious liquid was magically collected in a bucket or container. A thick ball of wool, containing a "spiritus," was rolled down the field and returned to its owner soaked in milk (Ferrer 2008). In a popular fable, an old woman sent her "spiritus" to look for butter, but the "tordyvel" stood in the path and did not know what to do or where to go. The old woman shouted irritably: "Du ä'en skit att dra!" (literally: You bring nothing but shit!). The "spiritus," who was a bit silly, in addition to being almost deaf, executed what he believed to be the order received, and the old woman was instantly buried in excrement (NMF n.d.).

Beetles were sometimes suspects of infernal conspiracy, and processed and accused of agricultural damage in the Court of Law. Particularly, *Melolontha* and weevils were held in contempt in several countries, including France, Switzerland, Sweden, and Norway (Somme 2004).

Ladybird (*Coccinella* L.) (a.k.a. ladybug or lady beetle) is a "bird," a holy "avian" (Ferrer 2009a), consecrated to Freja in northern Europe and much later Christianized to be one of the sacral attributes of the Virgin Mary. In medieval northern Europe, red was the color of love: the clothes of Freja, housewife of Frey, the *Venus Mater* of northern Europe, goddess of sexuality, fecundity, and prosperity, were symbolized by the ladybird (*Coccinella septempunctata*), a "goblin" creature.

The Swedish name for the seven-spot ladybird (Nyckelpigor) seems to be associated with the Virgin Mary ("Maria Nyckelpiga"), who according to tradition had the keys to the kingdom of heaven. In Catholicism, seven is a holy figure (i.e., the seven virtues of the Virgin Mary). The myth of Maria Nyckelpiga has been studied by the Swedish historian Backman (1947), who considered the name an exclusively Catholic legend. This author, however, never examined the old, Norse roots of the myth, ignoring fundamental aspects of the motive (Ferrer 2008, 2009a). In fact, the ladybird was demonized by Christianity and treated as a voracious, infernal creature. In the Gothic manuscript of *Beatus*, this characteristic beetle appears in a scene of hell, biting the foot of a sinner (Monserrat 2014). An iconographic analysis of the ladybird establishes the significance of the basic elements: a virginal maid is associated with a red, well-known beetle (*Coccinella septempunctata*). This association is due to a series of connotations, all of which are of sexual nature, because this beetle is extremely fertile: in northern Europe, after a cold winter, colonies of hundred individuals appear close to houses, surviving the ice and the snow. The beetles are extremely voracious, devouring harmful insects such as aphids, and for this reason are respected as good for the plants and the household in general. The beetle is red, the color of the goddess Freya, and the planet Mars. When disturbed, it secretes a yellow substance, historically associated with gold, because the terms yellow (*gul*) and golden (*guld*) were nearly homonyms in Nordic languages. As a more practical usage, these beetles were smashed and applied as anesthetic, in the case of dental infections (Tillhagen 1958). In fact these insects produce a peculiar alkaloid, close to xylocaine, which is an anesthetic. A flying ladybird was an omen of prosperity, and for this reason it was exhorted to stay, and then to fly again with various invocations, such as "Fly! Fly! come back carrying gold!" In Sweden, *Nyckelpiga* is also one of the names of Freya, as a representation of the lady of the house, administrating food and propriety, the domestic power symbolized by the key. The insect was previously treated as "Freya's Gulhöna" (Freya's golden hen).

The lock and the key are a very old sexual symbol, figured since 1450s in paintings (e.g., in *The Garden of Earthly Delights* by Hieronymus Bosch, and many others). The Swedish name "nyckelpiga" is intimately associated with the name "Peternycklas" (literally: Peter the key-man), a Swedish familiar name for the penis, and possibly a reference to St. Peter, known as the keeper of the gates of heaven, alluding to the port of the Paradise. Old paintings such as *The Unequal Couple* by Lucas Cranach the Elder (1530) seem to be iconographic allusions to the legends of Freya and ladybird (Ferrer 1984). Erwin Panofsky (1939) is probably correct in affirming that a painting (or any piece of artistic work) must be interpreted as a system of symbols reflecting the iconographic elements presented by the artist. Such analysis should be conducted in the context of the ideas available for the artist. In this sense, the painting not only has funny and moralizing aspects, but was probably a Protestant representation of St. Peter with the Virgin Mary. The personal symbols are clear: the Apostle has been figured as a bald, old man in every church facade, altarpiece, and painting. The red clothes of the girl correspond perfectly with the traditional attributes of the Virgin, but also of Freja and of the ladybird. This is not a surprising motif, painted by a German artist such as Lukas Cranach the Elder, who was a close friend of Luther. Naturally the Christian Church repudiated this kind of humor: both Catholics and Protestants tried to ban Freya and the irritating

"Golden bird," but failed to transform the "Maria Ladybird" into an attribute of the Virgin Mary. Some Swedish authors such as Backman (1947) interpreted this myth as a Catholic phenomenon that appeared as recently as during the reign of Queen Christina (around 1616). However, this interpretation is not plausible since representations of the Virgin Mary using red as the basic color of the mantel are present in older regions of central Europe, and for this reason it is perhaps one of the oldest myths in Scandinavia. The Italian and French *Madonnas* were normally represented with a blue mantel (except in the *Pietá*, clothed in black), but those in the Norse and German countries were draped in a red mantel. Since blue pigment *lapizlázuli* was too expensive or inaccessible in northern Europe, pigments obtained from pyrite, mercury, or lac were preferred in painting the celestial mantel of the Virgin.

In the remains of an old pharmacy from the fifteenth century found in archeological excavations in Stockholm, a jar containing smashed fragments of the golden beetle (*Cetonia aurata* L.) was found. It is hard to know why these insects were preserved in a jar in the pharmacy; perhaps to make ointment for cauterizing infected wounds and sores, or for sale as a fake aphrodisiac, since cantharids (*Lytta vesicatoria*) (Figure 8.4) are extremely rare insects in Sweden, contrarily to *Cetonia aurata*, which would not have worked in either of these cases. In another little jar a large number of graveyard beetles (*Rhizophagus parallelocollis*) (Coleoptera: Nitidulidae) were preserved. This is a curious beetle, whose larvae feed on human remains in graveyards and adults fly over the tombs. Another remarkable product was the powdered scarab beetles *Geotrupes*, used as a cheap laxative; and "canis album," dried and white dog excrement found after the long winter, used as white pigment in paintings (Tillhagen 1958).

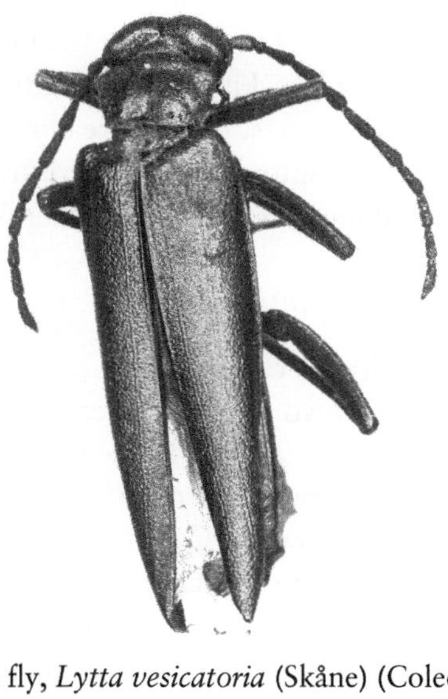

FIGURE 8.4 The Spanish fly, *Lytta vesicatoria* (Skåne) (Coleoptera: Cantharidae).

CANTHARIDS

The use and abuse of cantharids, known from Classical times, diminished in the Middle Ages as a result of the ignorance of old Greek and Latin sources. However, cantharids reappeared during the Renaissance with the translation and spread of some Classic works in Europe, such as those by Hippocrates (460–377 CE) and Dioscorides (30–90 CE), the first modern edition of which appeared during the Renaissance with the book of Mathias Ghast (1566). Cantharids have been used since the time of ancient cults of Dyonisus and Bacchus (*Bacchanales*) to achieve prolonged erections; in fact, cantharids provoked a long and dangerous erection (permanent priapism), resulting sometimes in fatal complications (gangrene). Unfortunately, the correct identification and knowledge of appropriate dosage for using these insects to prepare aphrodisiacs or to provoke abortions from old recipes was a serious and extremely dangerous obstacle. Accidents with poisoning and homicides were frequent (Hedin 1914). In northern Europe, Spanish flies (*Lytta vesicatoria*) were prosecuted by ecclesiastic and civil authorities and condemned in contumacy due to their failure to appear in court for the charges of conspiracy, convoked in Bern, Switzerland in 1475, and again in 1505 (Somme 2004; Ferrer 2014).

Cantharids as Symbols of Springtime

The Latin name *cantharis* was used in antiquity to refer to "dung beetles" (*Scarabaeus*, *Geotrupes*, etc.,) (Wessén 1966) and not to the modern cantharids, perhaps because a very old relationship exists between dung and fertility (Ferrer 2008). Linnaeus (1760) applied the name *cantharis* to another family of beetles, the Meloidae, exhibiting relatively soft tegument and extruding a peculiar substance, the cantharidine. Understanding some of the citations prior to Linnaeus is therefore not easy: for example, Pliny the Elder wrote that *Buprestis* can "kill an oxe" (*Historia Naturalis,* Liber 30, cap. 20: 3. 31) (Mayhoff 1906). Today buprestids are harmless insects, but a re-reading of Pliny's original text shows that what he describes is intoxication as a result of eating a cantharid beetle in a pasture, expressed more clearly in his Liber 37 (Latreille 1812).

Misinterpretations

Despite the notorious toxicity of cantharids known from antiquity (Cowen 1865; Scarborough 1977; O'Dowd 2001), in Germany these insects ("Käfer," cf. Hoffman-Krayer and Stäubl 1927–1942) were considered precursors of the springtime and solemnly carried in procession to the city. Grimm (1876) described how in some German towns, "Cantharid Käfer" were collected to be carried in procession, celebrating the arrival of the spring. In Sweden, cantharids are extremely rare and for this reason were confused by historians with other more common beetles, such as the melolonthids (*Melolontha*, *Amphimallon*, and *Rhizotrogus*), beetles that appear much later in the summer and cannot be collected at the beginning of the spring in Germany or Sweden. The great Swedish historian Backman (1947), who translated the German sources, interpreted these insects as "ollonborre," a diverse group of melolonthides. However, the German term is "cantharid," a completely different family of beetles. Backman was also unable to see the obvious relationship between ladybirds and the goddess Freja in northern Europe (Ferrer 2008). For this reason, it is possible to believe that he did not understand the symbolic and sexual role of cantharids as potent aphrodisiacs, interpreting the German processions simply as fertility rituals for the arrival of springtime. These German ceremonies were in fact vestiges of very ancient cults celebrating an exuberant sexuality, because cantharids were for a very long time used as aphrodisiacs in orgiastic festivals in honor of Bacchus in all regions submitted to Greek and Roman influence.

Lytta vesicatoria beetles sometimes erupt in massive concentrations of many thousand individuals, giving a bizarre forest concert. Cowen (1865) cites the case of a man badly intoxicated after sleeping close to a tree invaded by these beetles. Cantharids were used as a lethal ingredient in some infamous cases. Powdered cantharids were probably used by the poisoner *Locusta* in the Court of Nero and Tiberius to kill Britanicus, the son of Agrippina (Tacitus et al. 2008). Both *Meloe* and *Lytta* were named "cantharis" or cantharids, according to Suetonius Tranquillus (Suetonius 1957), who wrote: "Cantharis vermis est peresignum similis arachneae niger," a "bug" similar to a black spider, a comparison that cannot be applied to *Lytta* but fits much better with a *Meloe*.

The cutaneous use of cantharids is based on the authority of Pliny the Elder recommending cutaneous application against persistent sores: "Cantharides cum uva taminia intritae exedum, sed ita erosaa aliis, quae ad persuanda ulcera demostravimus, curari opotet" (Mayhoff 1906). According to authors such as Grieve (1971), *Uva taminia* probably refers to black bryony (*Tamus communis*). Pliny also described a tragic accident, resulting in the death of a Roman patient afflicted by *lupus* "liquen": despite the recommendation to use it as body lotion, the patient drank it and died. Several other tragic cases are also known (Blas 1953; Ferrer 2014).

The substance cantharidine was first scientifically obtained by M. Robiquet in 1810. Its posology is well known (Benjamin et al. 1940a, 1940b), however, it is extremely difficult to administer safely and without risk. Any dosage above 0.0002 grams (two-tenths of a milligram) can cause serious injury to the urinary and renal system, mucus membranes in general, and the urethra in particular, and a dose of 0.005 grams can be fatal.

Crickets *(Gryllus)*, singing continuously and everywhere, are a symbol of mental illness. In the illustrated medieval *Bestiare d'Amour* (Mermier 1992), Pierre de Beauvois represents a scary-looking cricket in a cage and writes, "The cricket so loses himself in his song that he sings himself to death" (Figure 8.5). A cricket is depicted by Hieronymus Bosch (1450–1516) as a grotesque

FIGURE 8.5 Cricket: Grillo or Krillon, the symbol of mental illness. *Bestiaire d'Amour*. MS Douce 308, f. 88v. Tracing ink of China drawing by the author.

FIGURE 8.6 Hieronymus Bosch, the Cricket-man. Detail from the triptych of the Temptations of St. Antonius, Museo national de Arte Antiga, Lisbon. Ink of China schematic drawing by the author.

cephalopodic human figure with very short legs (Figure 8.6). According to some specialists, this homunculus represents a cricket, tempting St. Anthony together with other demonic creatures in the desert.

Dragonflies and damselflies (*Odonata*) were sometimes used as decorative elements in illuminated manuscripts, but they were also considered infernal creatures with the capacity to cause a person to lose hair. In a very famous work, *Les Amants trépassés* (Cathedral of Strasbourg) painted at the end of the fourteenth century (Figure 8.7), a dragonfly is figured between other repulsive creatures, devouring the decayed body of the sinful women. The insect is figured so realistically that it is possible to identify it as a species belonging to the genus *Aeshna* (see Figures 8.6(c) to 8.7(f)). Another larger species irritating the woman in the painting is recognizable as a typical *Calopteryx* (see Figure 8.7(b)).

Flies (Figures 8.7(c)–8.7(e)) were sent by God as the fourth of the Biblical plagues against Egypt (Exodus 8:20–2). For this reason, flies were treated as satanic creatures, associated with the demon *Beelzebub*. In Norse mythology, the demonic god Loki transforms himself into a fly (Celander 1911). In some old Gothic paintings, a big black fly is figured flying in the mouth of a criminal sentenced to death (*Quentin Matsys*, Bruxelles). A big *Sarcophaga* fly is depicted close to the Madonna in some works of the Venetian artist Carlo Crivelli (1430–1495), suggesting that the figured woman is not Mary mother of Christ, but Mary Magdalene. These paintings, often interpreted as traditional representations of the Virgin Mary with Jesus, in fact seem to be heretic images of Mary Magdalene with the secret daughters of Christ, the *Saint Graal* according some heretic legends. The heresy of Cathars and other Christian sects probably inspired Crivelli to perform secret, heretic symbolic representations of a Madonna with child, adding flies in some paintings. The use of various symbols clearly associated with prostitution and sex during the medieval times supports this notion. Ferrer (2009a) indicated that this woman could not be the Virgin Mary, since yellow robes were reserved for prostitutes; the jewels decorating the crown may be *corona veneris* covering syphilitic eruptions; and because of the exuberant presentation of all obvious sexual and fertility symbols such as birds, apples, and flowers—e.g., pansies and clove, both associated with profane love, carried by this unconventional Madonna. For these reasons, this image is more likely to represent Mary Magdalene. Interestingly, a profanely ornate halo-crown around the head of the baby suggests that the child is in fact a girl.

Grasshoppers were one of the worst plagues of Egypt, sent to punish the tribes of Israel (Exodus 10:13–14, 19). Locusts were cited in the Apocalypse (9:3) as a massive infernal irruption, covering the face of the Earth. Curiously, Gregory the Great treated the grasshoppers as a symbol of the converted nations, combating the evil enemy. In the French Basilica of

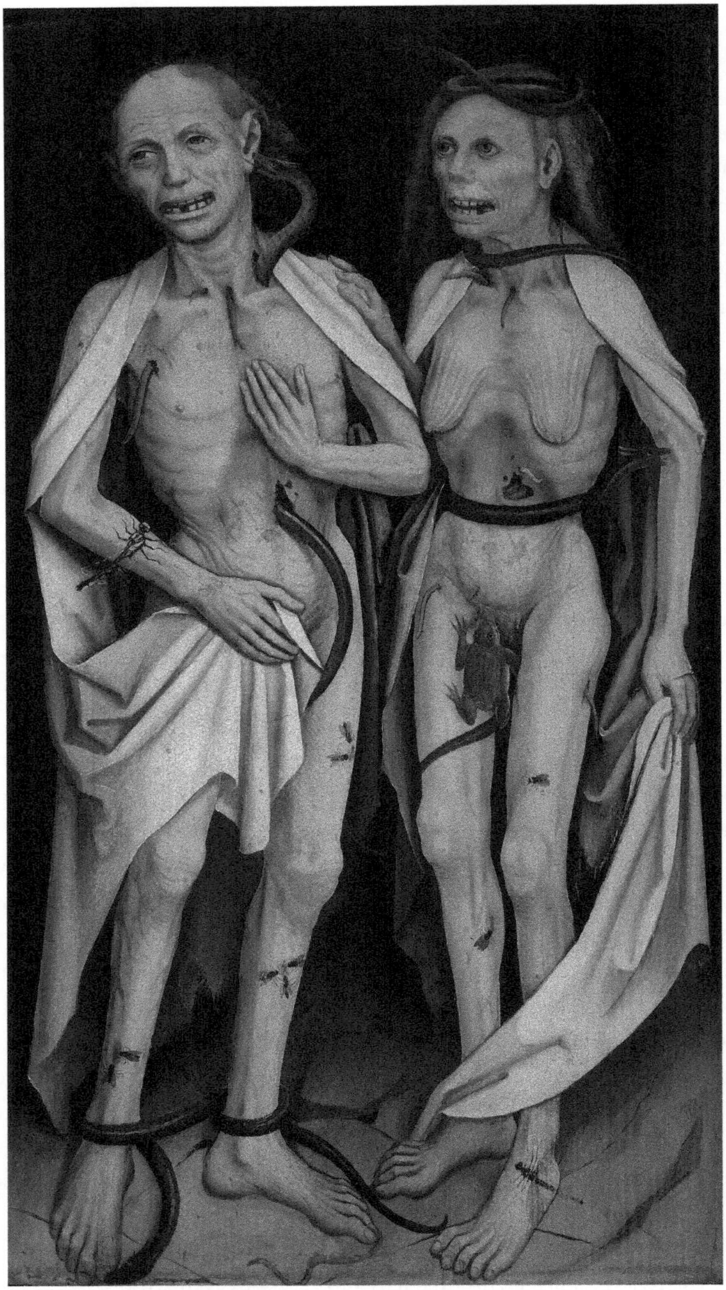

FIGURE 8.7 See caption on facing page

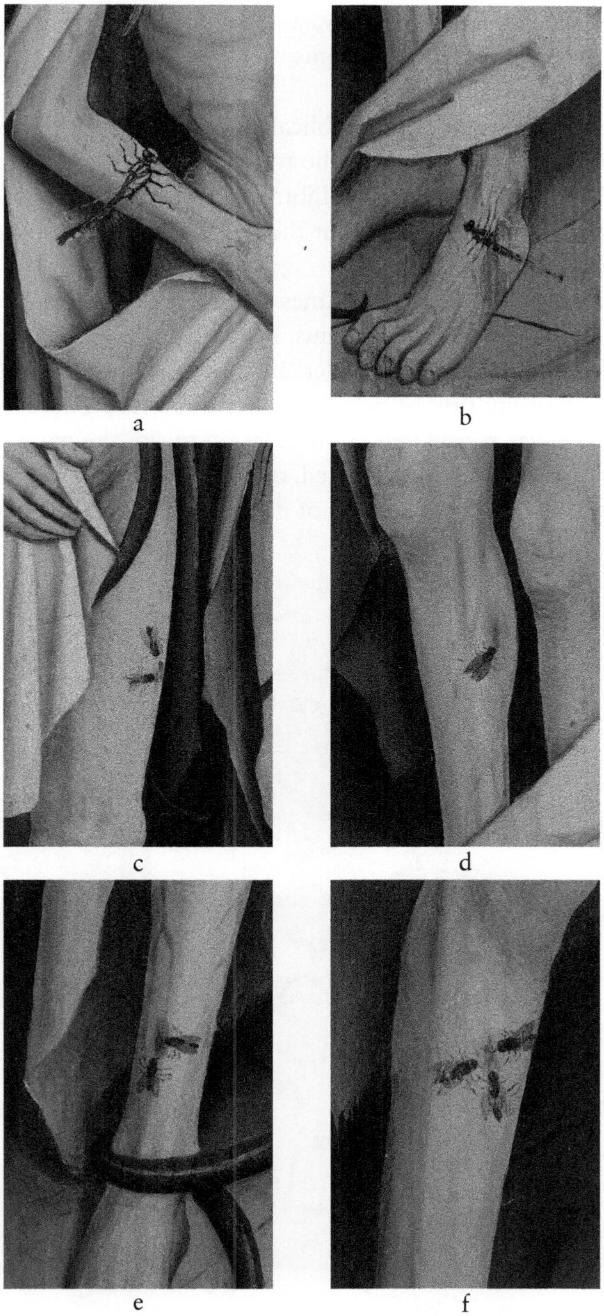

FIGURE 8.7 *Les Amants trépassés*. Strasbourg. Dragonflies, various flies, and worms figured among other repulsive creatures, devouring the decayed body of the sinful couple. Courtesy of Musée de l'Oeuvre de Notre Dame, Strasbourg. Detail: a) a large dragonfly irritating the man; b) a dragonfly on the woman's foot; c) two flies feeding on a lesion; d) a fly feeding on an open sore; e) two flies feeding on an oozing sore; f) three flies feeding on a lesion.

Vezelay from the thirteenth century, a sculpted grasshopper with a human head and grasshopper's body confronts a basilisk, probably representing this story (Cahier 1849) (Figure 8.8).

Maggots, according to many biblical sources, were represented as the inevitable, assiduous inhabitants of the tomb feeding on the corpse, as in the *Howard Psalter and Hours* (British Library 1310–1320) (Figure 8.9) and in several transi tombs (Figure 8.10), but the relationship of the maggot with the fly was probably unknown.

Mites. Parasites in general sometimes appear in Roman art as symbols of sin. A louse is destroyed by St. Sissinius, bishop of Reims, in a mural painting in the Cathedral of Reims (Beigbeder 1995). Later, when bathhouses were banned for their promotion of promiscuity (Figure 8.11), because nudity and corporal hygiene was considered as an occasion to sin, the natural parasites of the human body were not only accepted, but considered as a sign of welfare and prosperity, as in the commented case of dung beetles (Ferrer 2009b).

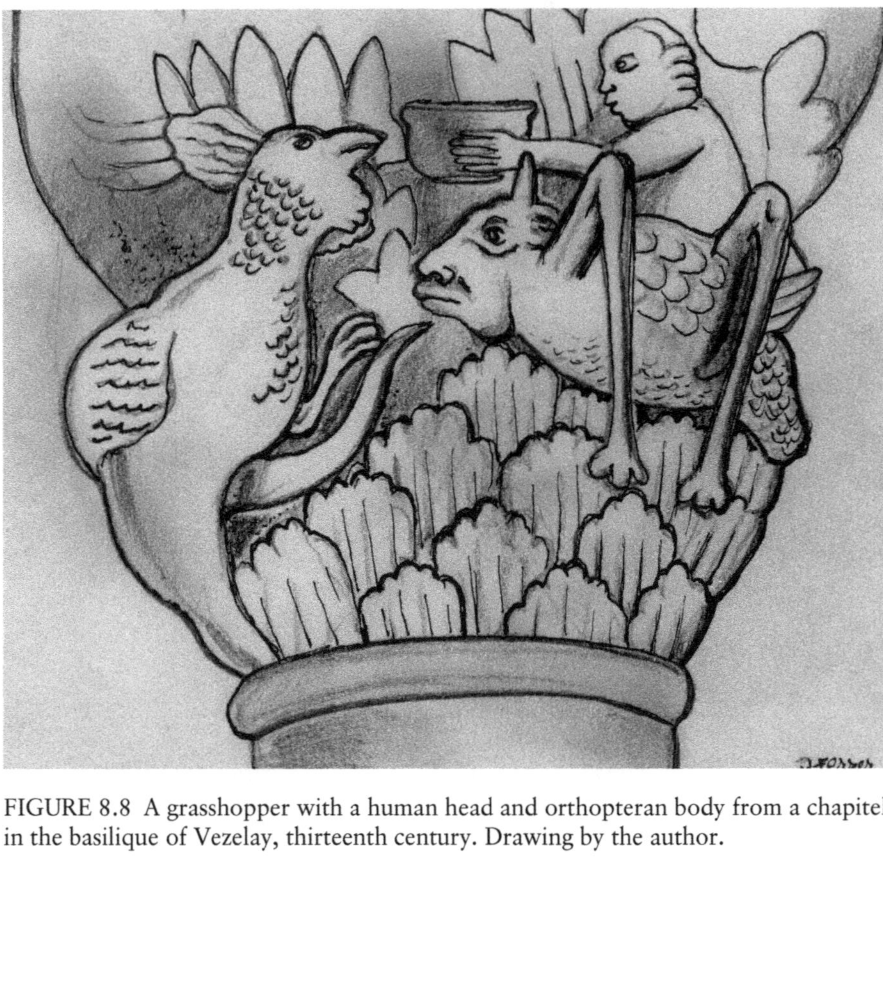

FIGURE 8.8 A grasshopper with a human head and orthopteran body from a chapitel in the basilique of Vezelay, thirteenth century. Drawing by the author.

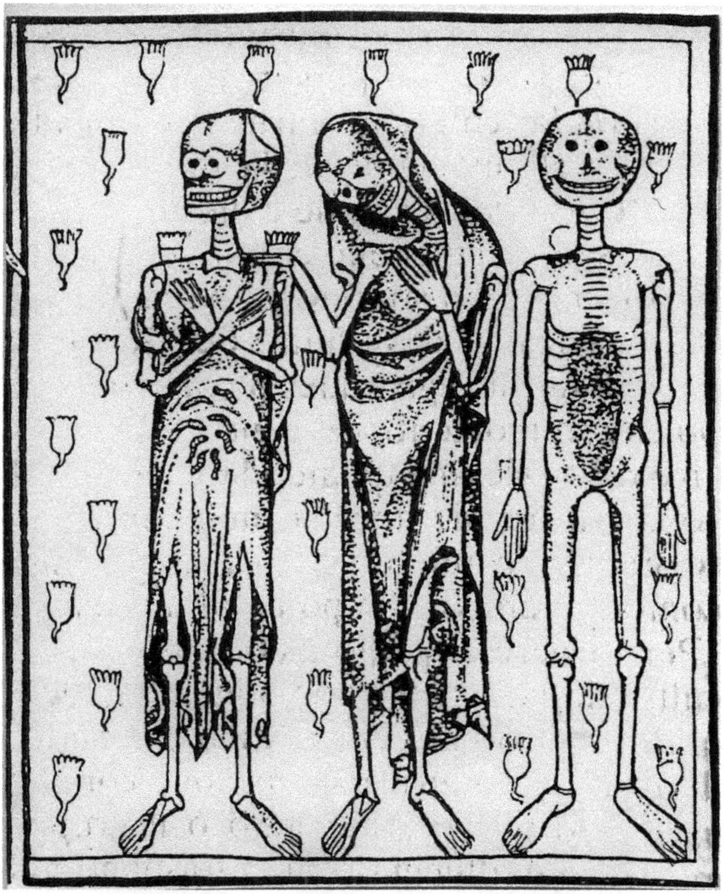

FIGURE 8.9 Detail from the *Howard Psalter and Hours* (thirteenth century). © British Library Board, Arundel MS 83, f. 127r PF.

Scorpions. As a result of the poisonous character of the scorpion, the animal, representing one of the signs of zodiac, was also considered as a symbol of evil and frequently associated with Jews in the Middle Ages. Astrological representations of scorpions as one of the signs of zodiac are usually accurate and realistic, but in religious texts the animal is often depicted in hideous and bizarre ways, similar to crabs or reptiles. In the Apocalypse, humans are tormented by terrible creatures with long hairy heads and with the body and wings of enormous locusts, armed like Horses of War: "And there came out of the smoke locusts upon the earth: and unto them was given power, as the scorpions of the earth have power" (Revelations 9:3), "And they had tails like

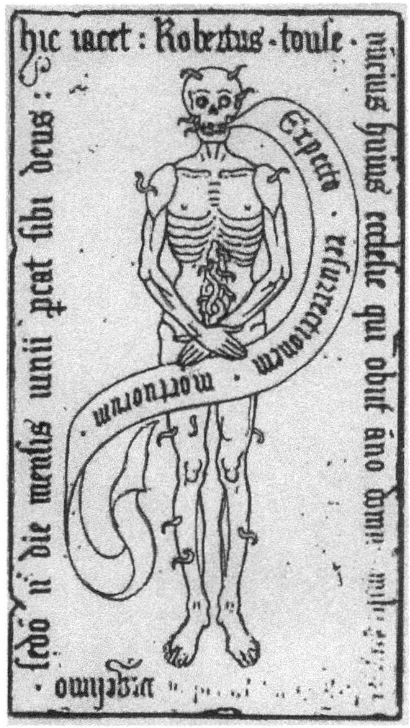

FIGURE 8.10 Waiting for the Resurrection Day; Tomb of Robertus Truse, Cathedral of Rouen, fourteenth century. Drawing by the author.

FIGURE 8.11 Anonymous, woodcut, Lovers in the Bath House, about 1480, Calendaire d'Ausbung. Courtesy of Wellcome Images.

unto scorpions, and there were stings in their tails: and their power was to hurt men five months" (Revelations 9:10).

Snails are not insects, but in the medieval Bestiaries and other books, they are sometimes figured as peculiar creatures with antennae similar to those of an insect. In the remarkable Book of Hours from the fourteenth century, a snail is represented with a cat head exhibiting antennae (Figure 8.12), while the Italian artist Carlo Crivelli placed a snail in a representation of the Virgin Mary (Figure 8.13). Perhaps the snail was chosen as a perfect symbol

FIGURE 8.12 Snail with a cat's head. Book of Hours, "*The Maastricht Hours*," Netherlands, early 1300. http://www.bl.uk/manuscripts/FullDisplay.aspx?ref=Stowe_MS_17. Tracing drawing by the author.

FIGURE 8.13 Snail, detail from *Carlo Crivelli: The Vierge and the Child with Saints Francis and Sebastian*, 1491. The National Gallery, London.

for the Cathars' doctrine because this animal was considered a sinner as it was found in vineyards. A knight fights against a suspiciously big snail in the mid-fourteenth-century Ormesby Psalter (Nazari 2014). The snail was rudely "purified" using salt and vinegar before the culinary treatment to be a delicious dinner.

Spiders were sometimes pictured with six legs as insects. In the Old Testament, a spider in a web protected King David when he was hiding in a cave; the enemies looking for him left after seeing the spider web, believing that the cave was empty. In Scandinavia, for a long time arachnids were considered mythical creatures associated with goblins and dwarfs, the jewelry masters and workers of subterranean mines. Curiously, the name for the spider is the same as "dwarf" in many Swedish dialects (Celander 1911). Killing spiders is frowned upon as their death can start rain; and their webs must also be protected. According to Isidore of Seville (*Etymologiae*, Book 12, 5:2), spiders are worms of the air. In a manuscript preserved in the Koninklijke Bibliotek, a spider exhibits six legs and a human face (see Figure 7.1). A much more realistic anthropomorphic depiction of a spider is found in the Italian manuscript *Seven Vices*, preserved in Genoa (1330–1340) (Figure 8.14).

FIGURE 8.14 Spider with anthropomorphic body, detail from a folio from the manuscript *Seven Vices*, Genoa, Italy, 1330–1340. © British Library Board, Add. manuscript 28841, f. 6. https://blogs.bl.uk/digitisedmanuscripts/2014/08/bugs-in-books.html.

FIGURE 8.15 Worms. Dance of Death, Totemdanz, about 1460, W. Stammler, Pinakotequè of Münich, Germany. Bequest of Herbert Mitchell, 2008, The Metropolitan Museum of Art.

FIGURE 8.16 Worms. Dance of Death, Totemdanz, about 1460, W. Stammler, Munich Digitization Center, Germany. Munich Digitization Center.

Worms were thought to be related to maggots in shape but were dissociated from flies, and just considered very big snakes (Figures 8.15 and 8.16). They were often figured devouring unburied corpses. The association with the snake in Paradise is clear in the iconography of death as the punishment of sin.

THE INSECTS ON THE CORPSES IN PAINTINGS FROM THE MIDDLE AGES TO THE RENAISSANCE

The painted skeletons in the beginning of the first half of the fifteenth century and the centuries prior are generally very poorly drawn, and in most cases it is impossible to recognize the species or sexes. According to Church doctrine, much like the angels, insects probably did not have sexes at all. However, the compositions are very sophisticated. One of the best-known examples is the *Howard Psalter and Hours* (1310–1320), with representations of three corpses in different stages of decomposition, one exhibiting the visceral cavity devoured by maggots. The larval stages of the flies feeding on the corpses show different phases: the first corpse on the left exhibits several maggots leaving the abdomen to pupate. The cadaver in the middle is covered with clothes and the insect activity is hidden. The last corpse on the right lacks genitals and clothes and the body is completely eviscerated by the larval mass of maggots.

By the end of the thirteenth and beginning of the fourteenth century, insects on human corpses were often accurately figured in several allegoric works as symbols of the punishment of the soul after death. One of the most representative works is *Les Amants trépassés* at the Musée de l'Oeuvre de Notre Dame, Strasbourg, France (Figure 8.7). For a long time this picture was attributed to Hans Baldung Grien, but modern research indicates that the artist was an anonymous master from the Upper Rhine. The artist paints different flies: some smaller houseflies depositing eggs in the injuries of the bodies (Figure 8.7(c)); a *Sarcophaga* easily identifiable by its size, grey body, and wings (Figure 8.7(d)); and remarkably, three flies appear, sucking blood from a wound on a living human (Figure 8.7(f)). The flies exhibit the characteristic projected mouthparts and ornate wings. Even more remarkable are the dragonflies (Odonata), so accurately drawn that identification to generic level is perfectly possible: they are a gaudy black and yellow *Aeshna* and a male of *Calipteryx* is recognizable by its purple wings (Figure 8.7(a)–(b)).

The history of the iconographic representation of death in the Middle Ages and the Renaissance indicate that for centuries, artists were very interested in the fate of human corpses, and in searching for realistic effects, their knowledge of the human anatomy become almost accurate long before advancements in medical sciences (Ferrer 2014). In paintings from the last decades of the fourteenth century, insects appear as accurate attributes of corpses, representing the infernal punishment of sin. Compared to earlier centuries, the interpretation of corpses painted during the Renaissance (fourteenth–sixteenth centuries) also become highly realistic and very different. A new aesthetic of death appeared in artistic manifestations. The postmortem status of the figured corpses suggests that the artists of this time used models, probably convicts killed during the early springtime. Realistic representations of dipteran larvae and other necrophagous insects use the intact epidermis as an umbrella for protection from the sun during the summer months while feeding from the flesh just under the corpse's skin.

INSECT PIGMENTS

Insect were a basic *materia prima* for medieval artists. In the Middle Ages, the Indian term *lac* ("hundred thousand") was applied to red pigment, extracted from a bug (*Laccifer laca*); the name referring to the great number of insects found on tree branches. These insects secreted a sticky resin which was preserved as a fine powder. This substance was boiled in urine with the addition of alum. Dried, the pigment was bound with a medium, often egg and oil in emulsion, and used to give brilliant color to sacred iconography in illuminated books (Ross 1971). A red pigment, *kermes*, was also obtained from dead, dried bodies of another insect, *Kermes vermilio*, to produce a beautiful sanguineous red pigment called *carmin* or *carmoisi*.

NOTES

Chapter 3

1. "… Similis cicadis vita, quarum duo genera: minores quae primae proveniunt et novissimae pereunt; sunt autem mutae. Sequens est volatura earum quae canunt; vocantur achetae et, quae minores ex his sunt, tettigonia. Sed illae magis canorae. Mares canunt in utroque genere, feminae silent" (Bostock and Riley 1855).
2. "… Quidam duo alia genera faciunt earum: surculariam, quae sit grandior; frumentariam, quam alii avenariam vocant, apparet enim simul cum frumentis arescentibus" (Bostock and Riley 1855).
3. "Deorum irae pestis ea intellegitur. namque et grandiores cernuntur et tanto volant pinnarum stridore, ut [aliae] alites credantur, solemque obumbrant sollicitis suspectantibus populis, ne suas operiant terras. Sufficiunt quippe vires et, tamquam parum sit maria transisse, inmensos tractus permeant diraque messibus nube contegunt, multa contactu adurentes, omnia vero morsu erodentes et fores quoque tectorum. Italiam ex Africa maxime coortae infestant… In Cyrenaica regione lex etiam est ter anno debellandi eas, primo ova obterendo, dein fetum, postremo adultas, desertoris poena in eum qui cessaverit… Graculos quoque ob id colunt adverso volatu occurrentes earum exitio. necare et in Syria militari imperio coguntur. Tot orbis partibus vagatur id malum" (Bostock and Riley 1855).
4. S. Agostino, *De Civitate Dei*, III, 31: One may also read that Africa, which had by that time become a province of Rome, was visited by a prodigious multitude of locusts, which, after consuming the fruit and foliage of the trees, were driven into the sea in one vast and measureless cloud; so that when they were drowned and cast upon the shore the air was polluted, and so serious a pestilence produced that in the kingdom of Masinissa alone they say there perished 800,000 persons, besides a much greater number in the neighbouring districts.
5. "Locustarum etiam in Africa multitudinem prodigii similem fuisse, cum iam esset populi Romani provincia, litteris mandaverunt; consumptis enim fructibus foliisque lignorum ingenti atque inaestimabili nube in mare dicunt esse deiectam;

qua mortua redditaque litoribus atque hinc aere corrupto tantam ortam pestilentiam, ut in solo regno Masinissae octingenta hominum milia perisse referantur et multo amplius in terris litoribus proximis" (literal translation by the authors).

Chapter 7

1. "insect, n.," "1601, P. Holland tr. Pliny *Hist. World* II. Catal. Words Art, '*Insects*, little vermine or small creatures, which have (as it were) a cut or diuision betweene their heads and bodies, as Pismires, Flies, Grashoppers, vnder which are comprehended Earthwormes, Caterpillers, [etc.]'" (*Oxford English Dictionary* 2018).
2. The original Latin text reads: "nusquam alibi spectatiore naturae rerum artificio," (Mayhoff 1906: 11.1). The English translation is by Eric C. Brown (2006: ix).
3. Perhaps most famously, in the Anglo-Saxon poem *Beowulf* the word "wyrm" is used on numerous occasions in reference to the dragon (Wrenn and Bolton 1988: 113, 183, 185, 193, 195, 196, 199, 201).
4. Hamlet then adds, "A certain convocation of politic worms are e'en at him. Your worm is the only emperor for diet: we fat all creatures else to fat us, and we fat ourselves for maggots" (Blakemore Evans 1997: 1219).
5. See Karl Steel on the female corpse's recognition of her "entanglement in an inescapably crowded, heterogeneous bodiment" of non-human life (Steel 2013: 102).
6. Beeswax was used as a writing surface on portable tablets, and it was also used to coat and seal manuscripts (Brown 1994: 7).
7. My thanks to Jane Wright for bringing this bee simile from *The Aeneid* to my attention.
8. The marginal note, translated by Andrew Galloway, reads: "his own humble John Gower, although long wearied in many ways by grave illness, did not refuse to take up the labours of this little work, but instead has most zealously compiled the present book from various chronicles, histories, and sayings of poets and philosophers, like a honeycomb gathered from various flowers" (Gastle 2010: 184).
9. See Steven Connor on "the significant etymological conversation between *atom* and *insect*" (2006: 90).
10. English translation by Amy King.
11. English translation by Amy King.

BIBLIOGRAPHY

Aberth, J. 2016. *An Environmental History of the Middle Ages: The Crucible of Nature*. London: Routledge.

Agger, W.A. and Maschner, H. 2009. "Medieval Norse and the Bidirectional Spread of Epidemic Disease Between Europe and Northeastern America: A New Hypothesis." In H. Maschner et al. (eds.), *The Northern World AD 900–1400*. Salt Lake City: The University of Utah Press.

Aiken, P. 1947. "The Animal History of Albertus Magnus and Thomas of Cantimpre." *Speculum* 22(2): 205–25.

Aldrovandi, U. 1602. *De animalibus insectis libri septem, cum singulorum iconibus ad vivum expressis*. Bologna: Giovan Battista Bellagamba.

Allcock, H. 1962. *Heraldic Design: It's Origins, Ancient Forms and Modern Usage*. New York: Tudor.

Allsop, K.A. and Miller, J.B. 1996. "Honey Revisited: A Reappraisal of Honey in Pre-Industrial Diets." *British Journal of Nutrition* 75: 513–20.

Alvarez, P.F.R. 1959. "Devoción popular a San Agustín en Andalucía en tiempos pasados." *Archivo Hispalense* 30: 197–208.

Ankarloo, B. 1971. *Trolldomsprocesserna i Sverige. Rättshistorisk Bibliotek, Institut för Rättshistorisk forskning*, bd. 17. Stockholm: Nordiska Bokhandel.

Anonymous. 1902. "Bee Hives in Battle." *British Bee Journal* 30: 466–7.

Anstey, M.L., Rogers, S.M., Ott, S.R., Burrows, M., and Simpson, S.J. 2009. "Serotonin Mediates Behavioural Gregarization Underlying Swarm Formation in Desert Locusts." *Science* 323: 627–30.

Araujo, A., Ferreira, L.F., Guidon, N., Maues Da Serra Freire, N., Reinhard, K.J., and Dittmar, K. 2000. "Ten Thousand Years of Head Lice Infection." *Parasitology Today* 16(7): 269.

Ariès, P. 1974. *Western Attitudes Towards Death From the Middle Ages to the Present*. Baltimore, MD: John Hopkins University Press.

Arrigo, N.C., Adams, P., and Weaver, S.C. 2010. "Evolutionary Patterns of Eastern Equine Encephalitis Virus in North Versus South America Suggest Ecological Differences and Taxonomic Revision." *Journal of Virology* 84(2): 1014–25.

Ashford, R.W.A. 2000. "Editorial: Parasites as Indicators of Human Biological Evolution." *Journal of Medical Microbiology* 49: 771–2.
Atchison, B. 2023. "Regulation of the Silk Industry – From the Book of the Eparch." Available online: https://www.pallasweb.com/deesis/byzantine-regulation-of-the-silk-industry.html#:~:text=To%20obtain%20admittance%20to%20the,and%20carry%20on%20a%20business (accessed May 26, 2023).
Atkins, E.L. 1948. "Mimicry Between the Drone-Fly, *Eristalis tenax* (L.), and the Honeybee, *Apis mellifera* L. Its Significance in Ancient Mythology and Present-Day Thought." *Annals of the Entomological Society of America* 41: 887–92.
Aulle, R. and Douteville, H. 1976. *Guide de la France Mistérieuse. Les Guides Noires.* Claude Tchou Editeur. Bologne et Paris.
Bachthold Stäubl, H. 1927–1942. *Handwörterbuch der des Deutschen Auberglaubens.* Gruyter Verlag, 10 vols.
Backman, L. 1947. *Jungfru Maria Nyckelpiga.* Stockholm: Norstedts.
Badke, D. 2004. *Commentary on St. Epiphanius on the Physiologus, by Christopher Plantin, 1588.* Available online: http://spcoll.library.uvic.ca/Digit/physiologum/animal/ant1.htm (accessed November 1, 2018).
Baekstedt, A. 1990. *Nordiska Gudar och Hjältar.* Oslo: När-Var Hur Serien, Forum.
Bain, A. 2004. "Irritating Intimates: The Archaeoentomology of Lice, Fleas and Bedbugs." *Northeast Historical Archaeology* 33: 81–90.
Balme, D.M., ed. and Peck, A.L., trans. 2014. *Aristotle, History of Animals*, Vol. 2. Cambridge, MA: Harvard University Press.
Barandarian, J.M. de. 1960. *Mitología Vasca.* Madrid: Biblioteca Vasca 5.
Barbattini, R. and Nicoli Aldini, R. 2010. "Le 'false api' nell'arte, nell'editorìa, nella pubblicità." *Apitalia* 10: 35–9, 11: 35–8.
Barber, R. and Riches, A. 1971. *A Dictionary of Fabulous Beasts.* London: Macmillan.
Barber, R., trans. 1992. *Bestiary: Being an English Version of the Bodleian Library, Oxford M.S. Bodley 764 with All the Original Miniatures Reproduced in Facsimile.* London: The Folio Society.
Barney, S.A., Lewis, W.J., Beach, J.A., and Berghof, O. 2006. *The Etymologies of Isidore of Seville.* New York: Cambridge University Press.
Bates, P. 2007. Sheep Scab (*Psoroptes ovis*). In I.D. Aitken (ed.), *Diseases of Sheep*, 321–5. Oxford: Oxford University Press.
Baur, O. 1974. *Bestiarium Humanum.* Munich: Heinz Moos Verlag.
Bayet, J. 1971. *Croyances et rites dans la Rome Antique.* Paris: Payot.
Beavis, I.C. 1988. *Insects and Other Invertebrates in Classical Antiquity.* Exeter, Devon: University of Exeter.
Beckwith, J. 1985. *Early Medieval Art: Carolingian, Ottonian, Romanesque.* London: Thames & Hudson.
Beckwith, J. 1997. *Arte Paleocristiano y Bizantino.* Madrid: Arte Cátedra.
Beigbeder, O. 1995. *Léxico de los Símbolos, Evocación del Arte Medieval.* Madrid: Encuentro.
Bell, D.N. 1992. *Wholly Animals: A Book of Beastly Tales.* Kalamazoo, MI: Cistercian Publications.
Ben-Ari, T., Neerinckx, S., Gage, K.L., Kreppel, K., Laudisoit, A., Herwig, L., and Stenseth, N.C. 2011. "Plague and Climate: Scales Matter." *PLoS Pathogens* 7(9): e1002160.
Benedictow, O.J. 2004. *The Black Death 1346–1352: The Complete History.* Woodbridge: Boydell Press.

Benito Del Caño, C. and Roldán Y Guerrero, R. 1928. *Cerámica farmacéutica, apuntes para su estudio*. Madrid: Jesús López.

Benson, L.D., ed. 2008. *The Riverside Chaucer*, 3rd edn. Oxford: Oxford University Press.

Bentley, M.T. 1988. "The Mimbres Butterfly Motif (The Rejuvenation of an Old Idea)." *The Artifact* 26(1): 39–79.

Benton, J.R. 1992. *The Medieval Menagerie, Animals in the Art of the Middle Ages*. New York: Abbeville Press.

Benton, J.R. 2002. *Art of the Middle Ages*. London: Thames & Hudson.

Berenbaum, M.R. 1995. *Bugs in the System: Insects and Their Impact on Human Affairs*. Helix Books, Addison-Wesley Publishing Company.

Berenbaum, M.R. 2018. "Bees, Ants, Spiders, and a Side of Bacon." *The Brooklyn Rail*. Available online: https://brooklynrail.org/2018/03/criticspage/Bees-Ants-Spiders-and-a-Side-of-Bacon (accessed November 1, 2018).

Berlo, J.C. 1983. "Warrior and the Butterfly: Central Mexican Ideologies of Sacred Warfare and Teotihuacan Iconography. In J.C. Berlo (ed.), *Text and Image in Pre-Columbian Art: Essays on the Interrelationship of the Verbal and Visual Arts*. Proceedings of the 44th International Congress of Americanists. International Series, vol. 180, 79–118. Oxford: British Archaeological Reports.

Berry, A.M. 1929. *Animals in Art*. London: Chatto & Windus.

Beullens, P. 2007. "Like a Book Written by God's Finger. Animals Showing the Path toward God." In B. Resl (ed.), *A Cultural History of Animals in the Medieval Age*, 127–51. Oxford and New York: Berg.

Bianciotto, G. 1980. *Bestiaires du Moyen Age*. Paris: Stock Plus.

Birdsong, R.E. 1934. "Insects of the Bible." *Bulletin of the Brooklyn Entomological Society* 29: 102–6.

Bishop, M. 1970. *The Middle Ages*. New York: American Heritage Press.

Bishop-Lilly, K.A., Ge, H., Butani, A., Osborne, B., Verratti, K., Mokashi, V., Nagarajan, N., Pop, M., Read, T.D., and Richards, A.L., 2013. "Genome Sequencing of Four Strains of Rickettsia Prowazekii, the Causative Agent of Epidemic Typhus, Including one Flying Squirrel Isolate." *Genome Announcements* 1(3): e00399–13.

Bitsch, C. 2013. *Natural Sciences Before the Letter: The Surprising Bestiary of Cocharelli*. Available online: https://www.museum.toulouse.fr/-/des-sciences-naturelles-avant-la-lettre-le-surprenant-bestiaire-des-cocharelli- (accessed May 16, 2018).

Black, J. and Green, A. 1992. *Gods, Demons and Symbols of Ancient Mesopotamia: An Illustrated Dictionary*. London: British Museum Press.

Blackham, H.J. 1985. *The Fable as Literature*. London: Athlone Press.

Blakemore Evans, G., ed. 1997. *Hamlet. In The Riverside Shakespeare*, 2nd edn. Boston and New York: Houghton Mifflin.

Blas, L. 1953. "Did King Ferdinand the Catholic Die From Drug Poisoning?" *Anales de la Real Academia de Farmacia* 19(2): 1011–108.

Blasco Zumeta, J. 1997. "Breve nota sobre langosta y superstición hasta la Ilustración del siglo XVIII." *Boletín de la Sociedad Entomológica Aragonesa* 20: 363–5.

Blatchford, C.H. 1889. "The Butterfly in Ancient Literature and Art." In S.H. Scudder (ed.), *The Butterflies of the Eastern United States and Georgia, with Special Reference to New England*, 3 vols., 1257–63. Cambridge, MA: published by the author.

Blázquez, J.M., Martínez-Pinna, J., and Montero, S., 2011. *Historia de las religiones antiguas: Oriente, Grecia y Roma*. Madrid: Cátedra.

Bodenheimer, F.S. 1947. "The Manna of Sinai." *Biblical Archaeology* 10: 2–6.
Bodenheimer, F.S. 1951. *Insects as Human Food: A Chapter of the Ecology of Man.* The Hague: Dr. W. Junk Publishers.
Bodleian Library. 2016. *Flemish Psalter MS. Douce 6.* Oxford, UK. Available online: http://bodley30.bodley.ox.ac.uk:8180/luna/servlet/detail/ODLodl~1~1~43633~120518:Psalter?trs=4&mi=0&qvq=q%3Abee+hive%3Blc%3AODLodl%7E14%7E14%2CODLodl%7E1%7E1%2CODLodl%7E23%7E23%2CODLodl%7E24%7E24%2CODLodl%7E6%7E6%2CODLodl%7E7%7E7%2CODLodl%7E8%7E8188dd33bbdba (accessed November 1, 2018).
Bos, K.I., Schuenemann, V.J., Golding, G.B., Burbano, H.A., Waglechner, N., Coombes, B.K., McPhee, J.B., Dewitte, S.N., Meyer, M., Schmedes, S., Wood, J., Earn, D.J., Herring, D.A., Bauer, P., Poinar, H.N., and Krause, J. 2011. "A Draft Genome of Yersinia Pestis From Victims of the Black Death." *Nature* 478: 506–10.
Bostock, J. and Riley, H.T. 1855. *The Natural History of Pliny*. London: H.G. Bohn. Available online: https://archive.org/details/naturalhistoryp00bostgoog/page/n4 (accessed November 1, 2018).
Boutellis, A., Abi-Rached, L., and Raoult, D. 2014. "The Origin and Distribution of Human Lice in the World." *Infection, Genetics and Evolution* 23: 209–17.
Bowers, B.S. 2007. *The Medieval Hospital and Medical Practice*. Burlington: Aldershot.
Brehaut, E. 1912. *An Encyclopedist of the Dark Ages: Isidore of Seville*. New York: Columbia University Press.
Brewer, J. and Sandved, K.B. 1976. "Butterflies in Art, Heraldry, and Religion." In *Butterflies*, 41–54. New York: H.N. Abrams.
Brion, M. 1959. *Animals in Art*. London: Harrap.
British Library. 1310–1320. *The Howard Psalter and Hours. Arundel 83 I.* Available online: http://www.bl.uk/catalogues/illuminatedmanuscripts/record.asp?MSID=6458&CollID=20&NStart=83 (accessed November 1, 2018).
British Library. 2018. *Bestiary, and Various Theological Texts. Royal 12 C XIX.* Available online: https://www.bl.uk/catalogues/illuminatedmanuscripts/ILLUMIN.ASP?Size=mid&IllID=55682 (accessed November 1, 2018).
Brown, E.C. 2006. "Introduction: Reading the Insect." In E.C. Brown (ed.), *Insect Poetics*. Minneapolis: University of Minnesota Press.
Brown, M.P. 1994. "The Role of the Wax Tablet in Medieval Literacy: A Reconsideration in Light of a Recent Finding From York." *British Library Journal* 20(1): 1–16.
Bruce, W.G. 1958. "Bible References to Insects and Other Arthropods." *Bulletin of the Entomological Society of America* 4(3): 75–8.
Bruce-Mitford, M. 2000. *El libro ilustrado de Signos y Símbolos*. Barcelona: Blume.
Brumble, H.D. 1973. "John Donne's 'The Flea': Some Implications of the Encyclopedic and Poetic Flea Traditions." *Critical Quarterly* 15(2): 147–54.
Bry, M. 2000. *Histoire des religions en Europe: judaïsmem, christianisme et Islam.* Paris: Hachette.
Buckland, P. and Sadler, J.P. 1989. "A Biogeography of the Human Flea, *Pulex irritans* L. (Siphoaptera: Pulicidae)." *Journal of Biogeography* 16: 115–20.
Buckland, P.C. 1981. "The Early Dispersal of Insect Pests of Stored Products as Indicated by Archaeological Records." *Journal of Stored Products Research* 17: 1–12.
Bulard, M. 1934. "Le Scorpion, symbole du peuple juif dans l'art religieux des XIVe, XVe, XVIe siècles." *Comptes rendus des séances de l'Académie des Inscriptions et Belles-Lettres* 78(1): 31–2.

Bulard, M. 1935. *Le scorpion. Symbole du people juif dans l'art religieux des XIVe, XVe, XVIe siécles*. Paris: De Boccard.

Büntgen, U. et al. 2011. "2500 Years of European Climate Variability and Human Susceptibility." *Science* 331: 578–58. DOI: 10.1126/science.1197175.

Buren, E.D. Van. 1937–1939. "The Scorpion in Mesopotamian Art and Religion." *Archiv für Orientforschung* 12: 1–28.

Buren, E.D. Van. 1939. *The Fauna of Ancient Mesopotamia. Analecta Orientalia 18*. Rome: Pontificium institutum biblicum.

Byrne, J.P. 2012. *Encyclopedia of the Black Death*, vol. 1. Santa Barbara, CA: ABC-CLIO.

Cahier, C. 1849. *Deux chapiteaux historiques du XIIéme siécle. Mélanges d'Archéologie, d'Histoire et de Litterature*, vol. 1. Paris: Poussielgue Rouscon ed.

Calkins, R.G. 1979. *Monuments of Medieval Art*. New York: E.P. Dutton.

Cambefort, Y. 1987. "Le scarabée dan l'Egypte ancienne: Origine et signification du symbole." *Revue d'Histoire des Religions* 204(1): 3–46.

Cambefort, Y. 1994a. *La Scarabéé et les dieux*. Paris: Boubée, Mondeville.

Cambefort, Y. 1994b. *Beetles as Religious Symbols*. Available online: https://www.insects.orkin.com/ced/issue-2/beetles-as-religious-symbols/ (accessed November 1, 2018).

Cambefort, Y. 2004. "Artistes, médecins et curieux aux origines de l'entomologie moderne." *Bulletin d'histoire et d'épistémologie des sciences de la vie* 11(1): 3–29.

Camille, M. 2000. *El ídolo gótico. Ideología y creación de imágenes en el Arte medieval*. Madrid: Akal.

Campagne, A. and Campagne, C. de la. 2005. *Animales extraños y fabulosos*. Madrid: Ed. Casariego.

Campbell, B. 2016. *The Great Transition: Climate, Disease and Society in the Late Medieval World*. Cambridge, MA: Cambridge University Press.

Camuffo, D. and Enzi, S. 1991. "Locust Invasions and Climactic Factors From the Middle Ages to 1800." *Theoretical and Applied Climatology* 43: 43–71.

Cantó Rubio, J. 1985. *Símbolos del arte cristiano*. Salamanca: Universidad Pontificia, Cátedra El Lenguaje del Arte.

Capatti, A. and Montanari, M., ed. 2003. *Italian Cuisine: A Cultural History*. Columbia University Press.

Caquot, A., Duchesmin-Guillemin, J., and Hadot, P. 1993. "Las religiones en el mundo mediterráneo y en el Oriente Próximo. Formación de las religiones universales y de salvación." In *Historia de las religiones*, vol. 5, *Formación de las religiones universales y de salvación – Las religiones en el mundo mediterráneo y en el Oriente Próximo*. Siglo XXI de España Editores, S.A. Madrid.

Carmona Muela, J. 1998. *Iconografía cristiana*. Madrid: Istmo.

Carmona Muela, J. 2000. *Iconografía clásica*. Madrid: Istmo.

Caro Y Cejudo, J.M. 1972. *Refranes, y modos de hablar castellanos: con los latinos que les corresponden, y la glosa, y explicación de los que tienen necesidad de ella: Con un indice de los adgios latinos, á los quales corresponden los castellanos, que van puestos en el libro por el órden de A. B. C.* Madrid: Imprenta Real.

Castelli, E. 2007. *Lo demoníaco en el arte: Su significado filosófico*. Madrid: Siruela.

Castillo, M. 1996. "Alberto Magno: Precursor de la Ciencia Renacentista." *La ciencia de los filósofos*: 91–106.

Cavallar, O., Dgenring, S., and Kirshner, J. 1994. *A Grammar of Signs: Bartolo da Sassoferrato's Tract on Insignia and Coats of Arms*. Berkeley, CA: Robbins Collection Publications.

Celander, H. 1911. "Lockes mytiska ursprung, Stockholm." *Språkvetenskapliga sällskapet i Uppsala, Förhandlingar* (1908–1909): 18–140.

Charbonneau-Lassay, L. 1996. *El bestiario de Cristo: el simbolismo animal en la Antigüedad y la Edad Media*. Madrid: José J. de Olañeta.

Chastel, A. 1984. *Musca depicta*. Milan: F.M. Ricci.

Chastel, A. and Manganelli, G. 1986. "A Fly in the Pigment: Iconography of the Fly in Paintings." *FMR* 19: 61–79.

Cherry, R. 1987. "History of Sericulture." *Bulletin of the Entomological Society of America* 33: 83–5.

Chevalier, J. and Gheerbrant, A. 1997. *Dictionnaire des symboles: Mythes, rêves, coutumes, gestes, formes, figures, couleurs, nombres*. Paris: Robert Laffont.

Child, H. 1971. *Christian Symbols, Ancient & Modern: A Handbook for Students*. London: Bell.

Cirlot, J.E. 1997. *Diccionario de símbolos*. Madrid: Siruela.

Clair, C. 1967. *Unnatural History: An Illustrated Bestiary*. New York: Abelard-Schumann.

Clark, W.B. 2006. *A Medieval Book of Beasts. The Second-Family Bestiary: Commentary, Art, Text and Translation*. Woodbridge: Boydell Press.

Clark, W.B. and McMunn, M.T. 1989. *Beasts and Birds of the Middle Ages: The Bestiary and its Legacy*. Philadelphia: University of Pennsylvania Press.

Clausen, L.W. 1954. *Insect Fact and Folklore*. New York: Macmillan.

Clauss, M. and Gordon, R. 2001. *The Roman Cult of Mithras: The God and His Mysteries*. Abingdon: Taylor & Francis.

Clébert, J.P. 1971. *Bestiaire Fabuleux*. Paris: Éditions Albin Michel.

Clemens, R. and Graham, T., eds. 2007. *Introduction to Manuscript Studies*. New York: Cornell University Press.

Cloudsley-Thompson, J.L. 1976. *Insects and History*. New York: St. Martin's.

Cloudsley-Thompson, J.L. 1990. "Scorpions in Mythology, Folklore and History." In G.A. Polis (ed.), *The Biology of Scorpions*, 462–85. Stanford University Press.

Cloudsley-Thompson, J.L. 2001. "Scorpions and Spiders in Mythology, Folklore and History." In V. Fet and P.A. Selden (eds.), *Scorpions 2001. In Memoriam Gary A. Polis*, 391–402. British Arachnological Society.

Cock-Starkey, C. 2016. "Strange Superstitions About 8 Everyday Insects From Around the World." *Mental Floss*. Available online: https://www.mentalfloss.com/article/86002/strange-superstitions-about-8-everyday-insects-around-world (accessed November 1, 2018).

Cohn, S.K. 2002. *The Black Death Transformed: Disease and Culture in Early Renaissance Europe*. London: Bloomsbury Academic.

Coimbra, C.E.A. 1988. "Human Settlements, Demographic Pattern, and Epidemiology in Lowland Amazonia: The Case of Chagas's Disease." *American Anthropologist* 90(1): 82–97.

Concina, C. 2016. "Unfolding the Cocharelli Codex: Some Preliminary Observations About the Text, with a Theory About the Order of the Fragments." *Medioevi* 2: 189–265.

Connor, S. 2006. *Fly*. London: Reaktion Books.

Contarino, R., ed. 1984. "La mosca." In L.B. Alberti et al. (eds.), *Apologhi ed elogi*. Genova: Edizioni Costa and Nolan.

Cooper, J.C. 2000. *Diccionario de símbolos*. Barcelona: Gustavo Gili.

Cooper, J.P. 1983. "Ideas of Gentility in Early Modern-England." In G.E. Aylmer and J.S. Morrill (eds.), *Men and Beliefs: Studies in Early-Modern History by J.P. Cooper*, 43–77. London: The Hambledon Pres.

Costa-Neto, E.M. 2000. "The Significance of the Category 'Insect' for Folk Biological Classification Systems." *Journal of Ecological Anthropology* 4: 70–5.

Costa-Neto, E.M. and Dunkel, F.V. 2016. Insects as Food: History, Culture and Modern Use Around the World. In A.T. Dossey et al. (eds.), *Insects as Sustainable Food Ingredients: Production, Processing and Food Applications*, 29–54. Cambridge, MA: Academic Press. Available online: http://dx.doi.org/10.1016/B978-0-12-802856-8.00002-8 (accessed April 27, 2023).

Cowan, F. 1865. *Curious facts of the History of Insects or Complete collection of the legends, superstitions, beliefs and ominous signs connected with Insects, together with their uses in medicine, art, and as Food and a Summary of their remarkable injuries and appearances*. Philadelphia: J.B. Lipping Ott and Co.

Cox, F. 2010. "History of the Discovery of the Malaria Parasites and Their Vectors." *Parasites and Vectors* 3: 1–9.

Crane, E. 1983. *The Archaeology of Beekeeping*. London: Duckworth.

Crane, E. 1999. *The World History of Beekeeping and Honey Hunting*. New York: Routledge and London: Duckworth.

Crane, E. and Graham, A.J. 1985. "Bee Hives of the Ancient World." *Bee World* 66(1): 23–41.

Crane, E. and Walker, P. 1999. "Early English Beekeeping: The Evidence From Local Records up to the End of the Norman Period." *The Local Historian* 29(3): 130–51.

Crapelet, G.A., ed. 1832. "La fourmi et le criquet." *Poésie morales et historiques d'Eustache Deschamp,* 191–2. Paris: Kessinger Publishing.

Crawford, D.H. 2007. *Deadly Companions: How Microbes Shaped our History*. Oxford: Oxford University Press.

Crescenzi, P. de. 1304–1309. *Liber ruralium commodorum*. Manuscript; first published 1471, [Augsburg]: Johannes Schüssler. Available online: http://bdh-rd.bne.es/viewer.vm?pid=d-1604508 (accessed November 1, 2018).

Crespo, F. and Lawrenz, M.B. 2014. "Heterogeneous Immunological Landscapes and Medieval Plague." In M.H. Green (ed.), *The Medieval Globe*, vol. 1, *Pandemic Disease in the Medieval World*, 229–58. Kalamazoo and Bradford: Arc Medieval Press.

Crocker, T.A. 2018. *Charlemagne's "Iron Crown."* Available online: https://www.charlemagne.org/ic.html (accessed April 1, 2018).

Crombie, A.C. 1952. "Master of the Cocharelli Codex, 'Cybo d'Hyeres: A Fourteenth-Century Zoological Artist'." *Endeavour* 11: 183–7.

Crombie, A.C. 1990. *Science, Optics, and Music in Medieval and Early Modern Thought*. London: The Hambledon Press.

Crombie, A.C. 2003. *The History of Science from Augustine to Galileo*. Mineola, NY: Dover Publications Inc.

Cui Y., Yu C., Yan Y., Li D., Li Y., Jombart T., Weinert L.A., Wang Z., Guo Z., Xu L., Zhang Y., Zheng H., Qin N., Xiao X., Wu M., Wang X., Zhou D., Qi Z., Du Z., Wu H., Yang X., Cao H., Wang H., Wang J., Yao S., Rakin A., Li Y., Falush D., Balloux F., Achtman M., Song Y., Wang J., and Yang R. 2013. "Historical Variations in Mutation Rate in an Epidemic Pathogen, *Yersinia Pestis.*" *Proceedings of the National Academy of Sciences of the United States of America* 110(2): 577–82.

Cumont, F. 1929. *Les Religions Orientales Dans Le Paganisme Romain*. Paris: Libraire Leroux. Available online: http://gallica.bnf.fr/ark:/12148/bpt6k5436659b/f380.image.r= (accessed November 1, 2018).

Curley, M.J., trans. 1979. *Physiologus*. Austin, TX and London: University of Texas Press.

Curley, M.J., trans. 2009. *Physiologus: A Medieval Book of Nature Lore*. Chicago and London: University of Chicago Press.

Czarski, C.M. 1999. "Kingdoms and Beasts: The Early Prophecies of Hildegard of Bingen." *Journal of Millennial Studies* 1(2): 1–10.

Dalby, A. 2011. *Geoponika – Farm Work: A Modern Translation of the Roman and Byzantine Farming Handbook*. London: Prospect Books.

Dalley, S. 1991. *Myths from Mesopotamia*. Oxford: Oxford University Press.

Dalley, S. 1998. *The Legacy of Mesopotamia*. Oxford: Clarendon Press.

Davies, M. and Kathirithamby, J. 1954. *Greek Insects*. London: Duckworth.

Davy, M.M. 1996. *Iniciación a la simbología románica: el siglo XII*. Madrid: Akal, Torrejón de Ardoz.

De Leemans, P. and Klemm, M. 2007. "Animals and Anthropology in Medieval Philosophy." In B. Resl (ed.), *A Cultural History of Animals in the Medieval Age*, 153–77. Oxford and New York: Berg.

Dean, J.M., ed. 2000. *Richard the Redeless and Mum and the Sothsegger*. TEAMS Middle English Texts Series. Kalamazoo, MI: Medieval Institute Publications. Available online: http://d.lib.rochester.edu/teams/text/dean-richard-the-redeless-and-mum-and-the-sothsegger-mum-and-the-sothsegger (accessed March 6, 2018).

Defoliart, G.R. 1989. "The Human Use of Insects as Food and as Animal Feed." *Bulletin of the Entomological Society of America* 35(1): 22–35.

Delaunay, P., Blanc, V., Del Giudice, P., Levy-Bencheton, A., Chosidow, O., Marty, P., and Brouqui, P. 2011. "Bedbugs and Infectious Diseases." *Clinical Infectious Diseases* 52(2): 200–10.

Dennys, R. 1975. *The Heraldic Imagination*. London: Barrie & Jenkins.

Di Nola, A.M. 2006. *La Nera Signora*. Antropologia della Morte ed dell lutto. Rome: Newton Comton eds.

Diekstra, F.N.M. 1985. "The Physiologus, The Bestiaries and Medieval Animal Lore." *Neophilologus* 69: 142–55.

Dittmar, K., Mamat, U., Whiting, M., Goldman, T., Reinhard, K., and Guillen, S. 2003. "Techniques of DNA-Studies on Pre-Hispanic Ectoparasites (Pulex sp., Pulicidae, Siphonaptera) from Animal Mummies of the Chiribaya Culture, Southern Peru." *Memóiras do Instituto Oswaldo Cruz* 98 (Suppl. 1): 53–8.

Dixon, B. 1996. "Ebola in Greece?" *British Medical Journal* 313: 430.

Dobson, M. 1980. "The Geography of Malaria in England." *Journal of Historical Geography* 6(4): 357–89.

Dobson, M. 2003. *Contours of Death and Disease in Early Modern England*, revised edn. Cambridge, MA: Cambridge University Press.

Dods, M., trans. 1950. *The City of God by Saint Augustine*. New York: The Modern Library, a division of Random House, Inc. Actual translation: 1872–1876.

Domínguez Rodríguez, A. 1993. *Iconografía de los libros de Horas del siglo XV de la Biblioteca Nacional*, 3 vols. Madrid: Universidad Complutense.

Donkin, R.A. 1977. "The Insect Dyes of Western and West-Central Asia." *Anthropos* 72: 847–80.

Dormer, S. 2012. *The Making of Medieval Illuminated Manuscripts' Transcript*. Available online: https://www.gresham.ac.uk/lectures-and-events/the-making-of-medieval-illuminated-manuscripts (accessed November 1, 2018).

Drancourt, M., Houhamdi, L., and Raoult, D. 2006. "Yersinia Pestis as a Telluric, Human-Ectoparasite Borne Organism." *The Lancet Infectious Diseases* 6(4): 231–41.

Druce, G.C. 1923. "[Myrmekoleon] or Ant-Lion." In *The Antiquaries Journal*, vol. III, No. 4, 347–63. London: Oxford University Press.

Duby, G. 1993. *A History of Private Life: Revelations of the Middle Ages,* vol. 2. Cambridge, MA: Belknap Press.

Durand, B., Lecollinet, S., Beck, C., Martinez-López, B., Balenghien, T., and Chevalier, V. 2013. "Identification of Hotspots in the European Union for the Introduction of Four Zoonotic Arborviroses by the Live Animal Trade." *PLOS One* 8(7): e7000, 1–16.

Duval, P.M. 1977. *Los Celtas, Universo de las Formas*. Madrid: Aguilar.

Dyer, T.R. 2013. *Some Notes On English Animal Lore – Birds, Animals, Insects And Reptiles (Folklore History Series)*. Kindle Edition. Pierides Press.

Eastaugh, N., Walsh, V., Chaplin, T., and Siddall, R., eds. 2004. *Pigment Compendium: A Dictionary of Historical Pigments*. Oxford and Burlington, MA: Elsevier.

Edwards, G.R. and Maupeu, P. 2015. *Le livre du pèlerin de vie humaine, 1355 / Guillaume de Deguileville, traduit [de l'ancien français] et présenté par Graham Robert Edwards et Philippe Maupeu*. Paris: Le Livre de Poche.

Edwards, T. 1908. *Lord of Honey Bee*. London: Methuen.

Edwards, T. 1920. *Bee – Master of Warrilow*. London: Methuen.

Eisen, R.J., Bearden, S.W., Wilder, A.P., Montenieri, J.A., Antolin, M.F., and Gage, K.L. 2006. "Early Phase Transmission of Yersinia Pestis by Unblocked Fleas as a Mechanism Explained Rapidly Spreading Plague Epizootics." *Proceedings of the National Academy of Sciences of the United States of America* 103(42): 15380–5.

Eisen, R.J., Petersen, J.M., Higgins, C.L., Wong, D., Levy, C.E., Mead, P.S., Schriefer, M.E., Griffith, K.S., Gage, K.L., and Beard, C.B. 2008. "Persistence of Yersinia Pestis in Soil Under Natural Conditions." *Emerging Infectious Diseases* 14(6): 941–3.

Eliade, M. 1999. *Historia de las creencias y de las ideas religiosas,* vol. 2, *De Gautama Buda al triunfo del cristianismo Paidós Ibérica*. Barcelona: Paidós.

Encyclopedia Britannica. 2023. "Penitential Book." Available online: https://www.britannica.com/topic/penitential-book (accessed May 24, 2023).

Encyclopedia of Fine Art. 2018. *Encaustic Painting*. Available online: http://www.visual-arts-cork.com/painting/encaustic.htm (accessed April 1, 2018).

Engel, M.S. and Kristensen, N.P. 2013. "A History of Entomological Classification." *Annual Review of Entomology* 58: 585–607.

Ettlinger, H.S. 1978. "The Virgin Snail." *Journal of the Warburg and Courtauld Institutes* 41: 316.

Evans, E.P. 1969. *Animal Symbolism in Ecclesiastical Architecture*. New York: Henry Holt and Co. Available online: https://archive.org/details/cu31924032663290 (accessed November 1, 2018).

Evans, G.R. 1983. *Alain of Lille: The Frontiers of Theology in the Late Twelfth Century*. Cambridge, MA: Cambridge University Press.

Fairclough, H.R., trans. 1994. *Eclogues, Georgics, Aeneid I–VI*. London: Harvard University Press.

Farson, D. and Hall, A. 1975. *Mysterious Monsters*. New York: Mayflower Books.

Ferguson, G. 1959. *Sings and Symbols in Christian Art*. Oxford: Oxford University Press.

Ferrer, J. 1984. *Moderna myter och gamla brott*. Stockholm: Norstedt & Söners förlag.

Ferrer, J. 2008. "Entomología fantástica: Los mitos nórdicos del escarabajo dorado, la mariquita y el estercolero." *Boletín de la Sociedad Entomológica Aragonesa* 43: 463–7.

Ferrer, J. 2009a. "Entomología fantástica: Las moscas de las Madonas de Carlo Crivelli." *Boletín de la Sociedad Entomológica Aragonesa* 44: 599–602.

Ferrer, J. 2009b. "Los mitos nórdicos del Escarabajo dorado, el Estercolero y la Mariquita." *Boletín de la Sociedad Entomológica Aragonesa* 43: 463–7.

Ferrer, J. 2014. "Entomología fantástica: Las moscas espanolas de Suecia." *Boletín de la Sociedad Entomológica Aragonesa* 51: 373–6.

Flandrin, J.L. Montanari, M., and Sonnenfeld, A., eds. 1999. *Food: A Culinary History from Antiquity to the Present*. Columbia University Press.

Flores, N.C. 1996. *Animals in the Middle Ages*. New York: Routledge.

Foottit, R.G. and Adler, P.H. 2016. *Insect Biodiversity Science and Society*, vol. II. Hoboken, NJ: Wiley-Blackwell.

Forbes, R.J. 1956. *Studies in Ancient Technology*, vol. IV. Leiden: E.J. Brill.

Forbes, V., Dussault, F., and Bain, A. 2005. "The 791 Equine Epidemic and its Impact on Charlemagne's Army." *Journal of Medieval Military History* 3: 23–45.

Forbes, V., Dussault, F., and Bain, A. 2013. "Contributions of Ectoparasite Studies in Archaeology with Two Examples From the North Atlantic Region." *International Journal of Palaeopathology* 3: 158–64.

Forrest, L.M. and Lasswell, J.L. 2005. *A Dazzle of Dragonflies*. Texas A&M University Press.

Fox-Davies, A.C. 1909. *A Complete Guide to Heraldry*. London: T.C. & E.C. Jack. Available online: https://archive.org/stream/completeguidetoh00foxdrich#page/n5/mode/2up (accessed November 1, 2018).

Fox-Davies, A.C. 1969. "Insects." In *A Complete Guide To Heraldry*, 195–6. London: Nelson & Sons.

Foys, M. 2018. "The Undoing of Exeter Book Riddle 47: 'Bookmoth'." In G.D. Caie and M.D.C. Drout (eds.), *Transitional States: Cultural Change, Tradition and Memory in Medieval England*. Tempe: Arizona Center for Medieval and Renaissance Studies (ACMRS), Humanities Commons. Available online: https://scholar.google.com/scholar?cluster=10884359817174460958&hl=en&as_sdt=0,36./ (accessed May 19, 2023).

Fraser, H.M. 1958. *History of Beekeeping in Britain*. London: Bee Res. Assoc.

Fraser, H.M. 1951. *Beekeeping in Antiquity*. London: University of London.

Freire García, L. 1995. "Las abejas de Sanson (Apis mellifica Samson, non L.)." *Boletín de la Sociedad Entomológica Aragonesa* 12: 31–2.

Friedmann, H. 1980. *A Bestiary for Saint Jerome: Animal Symbolism in European Religious Art*. Washington, DC: Smithsonian Institution.

Frost, S.W. 1937. "The Insects Motif in Art." *The Scientific Monthly* 44(1): 77–83.

Gagliardi, R.A. 1976. *The Butterfly and Moth as Symbols in Western Art*. MSc diss, Southern Connecticut State College, New Haven. Available online: https://www.proquest.com/openview/b894d241e473fbd889b6658cf47702c8/1?pq-origsite=gscholar&cbl=18750&diss=y (accessed November 1, 2018).

Gagliardi, R.A. 1997. "Lepidoptera Symbols Relating to Wings and the Body." *Cultural Entomology Digest* 10: 53.

Gagliardi, R.A. 1996. *The Butterfly and Moth as Symbols in Western Art*. MS thesis, South Conneticut, State College, New Haven.

Gallo, A. 1569. *Le vinti giornate dell'agricoltura e i piaceri della villa*. Venice: Grazioso Peraccino.
Gastle, B. 2010. "Gower's Business: Artistic Production of Cultural Capital and the Tale of Florient." In E. Dutton et al. (eds.), *John Gower, Trilingual Poet: Language, Translation, and Traditio*, 182–95. Cambridge: D.S. Brewer.
Geraghty, S. 1996. *Viking Dublin: Botanical Evidence from Fishamble Street. Medieval Dublin Excavations 1962–1981: Series C*, vol. 2. Dublin: Royal Irish Academy.
Gerhardt, M. 1965. "The Ant-Lion, Nature Study on the Interpretation of a Biblical Text From the Physyologus to Albert Le Great." *Vivarium* 3: 1–23.
Getz, F.M. 1998. *Medicine in the English Middle Ages*. Princeton, NJ: Princeton University Press.
Ghast, M. 1566. *Pedanius Dioscórides, De Materia Medica*. In folio (28). Salamanca.
Gibb, H.A.R., trans. 1929. *Ibn Battúta: Travels in Asia and Africa, 1325–1354*. London: Routledge & Kegan Paul.
Gigante, M., trans. 1964. *Vituperio Della Mosca. Eugenii Panormitani versus iambici*. Palermo: Istituto Siciliano di Studi Bizantini e Neoellenici.
Gillmor, C. 2005. "The 791 Equine Epidemic and its Impact on Charlemagne's Army." *The Journal of Medieval Military History*. The Boydell Press III: 23–45.
Gimbutas, M. 1991. *Diosas y dioses de la Vieja Europa 7000–3500 A.C.* Mitos, leyendas e imaginería. Madrid: Istmo.
Gimbutas, M. 1996. *El lenguaje de la Diosa*. Oviedo: GEA.
Glick, T.F., Livesey, S.J., and Wallis, F., eds. 2005. "Use of Insects." *Medieval Science, Technology, and Medicine: An Encyclopedia*. New York and London: Routledge.
Goor, A. 1965. "The History of the Fig in the Holy Land from Ancient Times to the Present." *Economic Botany* 19: 124–35.
Gottfried, R.S. 1983. *The Black Death: Natural and Human Disaster in Medieval Europe*. New York: The Free Press.
Gough, M. 1973. *The Origins of Christian Art*. London: Thames & Hudson.
Grabar, A. 1980. *Christian Iconography: A Study of its Origins*. Princeton, NJ: Princeton University.
Grimaldi, D. and Engel, M.S. 2005. *Evolution of the Insects*. Cambridge, MA: Cambridge University Press.
Grant, R.M. 1999. *Early Christians and Animals*. London: Routledge.
Graves, R. 1989. *The Twelve Caesars*. London: Harmondsworth, Penguin Classics.
Graves, R. 1996. *Los Mitos Griegos*, 2 vols. Madrid: Alianza Editorial.
Grieve, M. 1971. *A Modern Herbal*, 2 vols. Ed. C.F. Leyel. Mineola, NY: Dover Publications Inc.
Grimm, J. 1876. *Deutsche Mythologie*, 4th edn. Curated Elard Hugo Meyer. Berlin: F. Dümmler, 1875–1878, 3 vols. Reprinted Darmstadt: Wissenschaftliche Buchgesellschaft, 1965.
Grinnell, G.B. 1899. "The Butterfly and the Spider Among the Blackfeet." *American Anthropologist* (n.s.): 194–6.
Gross, L. 1995. "How the Plague Bacillus and its Transmission Through Fleas Were Discovered: Reminiscences From My Years at the Pasteur Institute in Paris." *Proceedings of the National Academy of Sciences of the United States of America* 92(17): 7609–11.
Gubernatis, A. de. 1968. *Mythology; or The Legends of Animals*, 2 vols. Detroit, MI: Singing Tree Press.
Gubler, D.J. 2009. "Vector-Borne Diseases." *Revue Scientifique et Technique de l'OIE* 28: 583–8.

Guerra, M. 1986. *Simbología románica: el cristianismo y otras religiones en el arte románico*. Madrid: Fundación Universitaria española.

Gugliuzzo, E. and Restifo, G. 2014. *La piaga delle locuste. Ambiente e società nel Mediterraneo d'età moderna*. Naples: Giapeto editore.

Haensch, S., Bianucci, R., Signoli, M., Rajerison, M., Schultz, M., Kacki, S., Vermunt, M., Weston, D.A., Hurst, D., Achtman, M., Carniel, E., and Bramanti, B. 2010. "Distinct Clones of Yersinia Pestis Caused the Black Death." *PLoS Pathogens* 6(10): e1001134.

Hagen, H.A. 1884. "Lawsuits Against Grubs and Grasshoppers." *Science* 4(82): 168–71.

Hale, J.R. 1998. "The Viking Longship." *Scientific American* February 1: 56–63.

Hall, J. 1974. *Diccionario de temas y símbolos artísticos*. Madrid: Alianza.

Hallman, M.S. 1951. "The Story of Honey Bees." *Bios* 22(3): 198–208.

Halsall, P. 1998. *Medieval Sourcebook: Procopius: The Roman Silk Industry, c. 550*. Available online: https://sourcebooks.fordham.edu/source/550byzsilk.asp (accessed April 1, 2018).

Harbeck, M., Seifert, L., Hänsch, S., Wagner, D.M., Birdsell, D., Parise, K.L., Wiechmann, I., Grupe, G., Thomas, A., Keim, P., Zöller, L., Bramanti, B., Riehm, J.M., and Scholz, H.C. 2013. "Yersinia Pestis DNA from Skeletal Remains from the 6th Century AD Reveals Insights into Justinianic Plague." *PLoS Pathogens* 9(5): e1003349.

Hassig, D. 1995. *Medieval Bestiaries: Text, Image, Ideology*. New York: Cambridge University Press.

Hatcher, P. and Battey, N. 2011. *Biological Diversity: Exploiters and Exploited*. John Wiley & Sons.

Hawthorne, J.G. and Smith, C.S. 1963. *On Divers Arts: The Treatise of Theophilus*. Chicago, IL: University of Chicago Press.

Hays, H.R. 1973. *Birds, Beasts, and Men: A Humanist History of Zoology*. Baltimore, MD: Penguin Books.

Headrick, A. 2003. "Butterfly War at Teotihuacan." In M.K. Brown and T.W. Stanton (eds.), *Ancient Mesoamerican* Warfare, 149–70. Walnut Creek, CA: AltaMira Press.

Hecht, B.P. and Parks, L.M. 1940a. "Studies on Cantharides. 1. The Titration of Cantharidin." *Journal of the American Pharmaceutical Association* 29(2): 71–7.

Hecht, B.P., & Parks, L.M. 1940b. Studies on Cantharides. 2. The Essay of Cantharides. *Journal of the American Pharmaceutical Association*, 29(3): 111–15.

Heck, C. and Cordonnier, R. 2012. *The Grand Medieval Bestiary: Animals in Illuminated Manuscripts*. New York: Abbeville Press.

Hedin, A. 1914. *Giftmord och Djävulsmässor, giftmordsfarsoten i Frankrike på Ludvigs XIVs tid. Ur författarens efterlämnade papers*. Stockholm and Göteborg: Åhlen & Åkerlund förlag.

Heider, G.C. 1995. *Dictionary of Deities and Demons in the Bible*. Leiden: E.J. Brill.

Henderson, A.C. 1982. "Medieval Beasts and Modern Cages: The Making of Meaning in Fables and Bestiaries." *Publications of the Modern Languages Association of America* 97: 40–9.

Henderson, J., ed. 1913. "The Fly*The Works of Lucian*, vol. 1. Cambridge, MA and London: Loeb Classical Library. Available online: https://www-loebclassics-com.bris.idm.oclc.org/view/lucian-fly/1913/pb_LCL014.83.xml (accessed March 7, 2018).

Herrod-Hempsall, W. 1937. *Bee-Keeping New and Old: Described with Pen and Camera*, vol. II. London: The British Bee Journal.

Hersig, D. 1995. *Medieval Bestiaries, Text, Image, Ideology*. Cambridge, MA: Cambridge University Press.
Heyden, D. and Baus Czitrom, C. 1991. "Los insectos en el arte prehispánico." *Artes de México, revista-libro "Insectos y artrópodos en el arte mexicano"* (2nd edn., 1997): 25–37.
Hicks, C. 1993. *Animals in Early Medieval Art*. Edinburgh: Edinburgh University Press.
Hinnebusch, B.J. 2005. "The Evolution of Flea-Borne Transmission in *Yersinia Pestis*." *Current Issues in Molecular Biology* 7: 197–212.
Hoffman Krayer, E. and Bachthold Stäubl, H. 1927–1942. *Handwörterbuch der des Deutschen Auberglaubens*, vol. 5, *Folklore and Mythology*, 10 vols. Walter de Gruyter Verlag.
Hoffmann, R.C. 2014. *An Environmental History of Medieval Europe*. Cambridge, MA: Cambridge University Press.
Hoffmann, R.C. 2010. "Bugs, Beasts and Business: Some Everyday and Long-Term Interaction Between Biology and Economy in Preindustrial Europe." In S. Cavaciocchi (ed.), *Le interazioni fra economia e ambiente biologico nell'Europa preindustriale, secc. XIII–XVIII*. Florence: University Press.
Hogue, C.L. 1975. "The Insect in Human Symbolism." *Terra* 13(3): 3–9.
Hogue, C.L. 1987. "Cultural Entomology." *Annual Review of Entomology* 32: 181–99.
Hogue, J.N. 2009. "Chapter 101: Folk Beliefs and Superstitions." In V.H. Resh and R.T. Cardé (eds.), *Encyclopedia of Insects*, 372–6. Elsevier, Academic Press. Available online: https://doi.org/10.1016/B978-0-12-374144-8.00110-7 (accessed April 28, 2023).
Hope, S. 2013. *The Humility of Snails*. My Albion Blog, part 1: September, part 2: October. Available online: http://my-albion.blogspot.no (accessed November 1, 2018).
Hopla, C.E., Durden, L.A., and Keirans, J.E. 1994. "Ectoparasites and Classification." *Revue Scientifique et Technique de l'OIE* 13(4): 985–1017.
Horstmann, C., ed. 1851. *The Bee and the Stork. MS Thornton (Lincoln Cathedral 91), Folio 194*. English prose works of Richard Rolle: A selection. Corpus of Middle English Prose and Verse. Available online: https://quod.lib.umich.edu/c/cme/rollewks/1:1.5.1?rgn=div3;view=fulltext (accessed March 6, 2018).
Houston, K. 2016. *The Book: A Cover-to-Cover Exploration of the Most Powerful Object of Our Time*. New York: W.W. Norton & Company.
Houwen, L.A.J.R. 1997. *Animals and the Symbolic in Mediaeval Art and Literature*. Groningen: Egbert Forsten.
Howard, L.O. 1895. "A Brief Account of the Rise and Present Condition of Official Economic Entomology." In *Insect Life*, vol. VII, 55–108. US Department of Agriculture, Division of Entomology, Periodical Bulletin. Washington.
Huchet, J.-B. 2015. "Ectoparasites de l'homme." In F. Collard and E. Samama (eds.), *Poux, Puces, Punaises la Vermine de l'Homme: Découverte, descriptions et traitements Antiquité, Moyen Âge, Époque moderne*, 45–60. Paris: l'Harmattan.
Hunt, L.B. 1980. "The Long History of Lost Wax Casting." *Gold Bulletin* 13(2): 63–79.
Hunt, P. 2007. *Byzantine Silk: Smuggling and Espionage in the 6th Century CE*. Available online: http://altmarius.ning.com/profiles/blogs/byzantine-silk-smuggling-and-espionage-in-the-6th-century-ce (accessed November 1, 2018).

Hutchinson, G.E. 1974. "Marginalia: Aposematic Insects and the Master of the Brussels Initials: The Late Medieval Tradition of Iconography of Nature, Exemplified in the Work of this 15th-Century Manuscript Painter, Contributed to the Rise of Modern Biology." *American Scientist* 62(2): 161–71.

Hyland, A. 1994. *The Medieval Warhorse, from Byzantium to the Crusades*. Stroud: Sutton Publishing Ltd.

Impelluso, L. 2003. *La Naturaleza y sus símbolos*. Milan: Electa.

Isidoro de Sevilla. 2004. *Etymologiae*. Madrid: Biblioteca de authoires Cristianos.

Izquierdo, I. and Le Meaux, H. 2002. *Seres Híbridos*. Madrid: Museo Arqueológico Nacional.

Jacobs, J., ed. 1985. *The Fables of Odo of Cheriton*. New York: Syracuse University Press.

Jacobs, J., ed. 1889. "The Ant and the Sygale." *The Fables of Aesop, as first printed by William Caxton in 1484*. London: David Nutt.

Jones, C.W., ed. 2013. "The Voice of the Turtle." *Medieval Literature in Translation*, 258. New York: Dover Books.

Jones, E.J. 1943. *Medieval Heraldry: Some Fourteenth Century Heraldic Works*. Cardiff: William Lewis, Ltd.

Jones, P.M. 1998. *Medieval Medicine in Illuminated Manuscripts*. London: The British Library.

Jones, R. 2013. *The Medieval Natural World*. Harlow: Pearson Education Ltd.

Keel, O. and Uehlinger, C. 1998. *Gods, Goddesses, and Images of God*. London: Bloomsbury Academic.

Keiser, I. 1966. "Insects and Related Arthropods in Heraldry." *Bulletin of the Entomological Society of America* 12: 314–18.

Kenward, H. and Allison, E. 1994. "Rural Origins of the Urban Insect Fauna." In A. Hall and H. Kenward (eds.), *Urban-Rural Connexions: Perspectives from Environmental Archaeology*, 55–79. Symposia of the Association of Environmental Archaeology No. 12. Oxford: Oxbow Books.

Kenward, H., Hall, A., Allison, E., and Carrott, J. 2011. "Environment, Activity and Living Conditions at Deer Park Farms: Evidence from Plant and Invertebrate Remains." In C. Lynn and J. McDowell (eds.), *Deer Park Farms: The Excavation of a Raised Rath in the Glenarm Valley, Co Antrim*, 497–548. Antrim: Stationary Office and Environment and Heritage Service.

Kern, D. 2005. "Les Libellules des manuscrits enluminés du Moyen Age." *Martinia* 21: 35–42.

Kevan, D.K. McE. 1992. "Antlion ante Linné: [Myrmekoleon] to Myrmeleon (Insecta: Neuroptera: Myrmeleonidae [sic]." In M. Canard et al. (eds.), *Current Research in Neuropterology*. Proceedings of the Fourth International Symposium on Neuropterology, 203–32. [Symposium held in Bagnères-de-Luchon, France, 1991.] Toulouse: M. Canard.

Khateeb Shehada, H.A. 2013. *Mamluks and Animals: Veterinary Medicine in Medieval Islam*. Leiden: E.J. Brill.

King, G. and Henderson, C. 2014. "Living Cheek By Jowl: The Pathoecology of Medieval York." *Quaternary International* 341: 131–42.

King, G., Kenward, H., Schmidt, E., and Smith, D. 2014. "Six-Legged Hitchhikers: An Archaeobiogeographical Account of the Early Dispersal of Grain Beetles." *Journal of the North Atlantic* 23(1): 1–18.

Kitchell, K. and Resnick, M. 1999. *Albertus Magnus, On Animals: A Medieval Summa Zoologica*. Baltimore and London: The Johns Hopkins University Press.
Kliffen, I. 1996. *Celtic Animals Charted Designs*. Mineola, NY: Dover Publications Inc.
Klingender, F. 1971. *Animals in Art and Thought to the End of the Middle Ages*. London: Routledge & Kegan.
Krapp, G.P. and Dobbie, E.V.K., eds. 1936. *The Exeter Book: The Anglo-Saxon Poetic Records*, vol. 3. New York: Columbia University Press.
Kritsky, G. 1993. "Beetle Gods, King Bees and Other Insects of Ancient Egypt." *KTM*, 4: 32–9.
Kritsky, G. 1997. "The Insects and Other Arthropods of the Bible, the New Revised Version." *American Entomologist* 43(3): 183–8.
Kritsky, G. 2010. *The Quest for the Perfect Hive*. New York: Oxford University Press.
Kritsky, G. 2015. *The Tears of Re: Beekeeping in Ancient Egypt*. New York: Oxford University Press.
Kritsky, G. 2016. "A New Beekeeping Exhibit at the Musée de l'Homme in Paris." *American Bee Journal* 156: 1025–8.
Kritsky, G. 2017. "Beekeeping from Antiquity through the Middle Ages." *Annual Review of Entomology* 62: 249–64.
Kritsky, G. and Cherry, R. 2000. *Insect Mythology*. San Jose, CA: Writers Club Press.
Lans, C., Turner, N., and Khan, T. 2008. "Medicinal Plant Treatments for Fleas and Ear Problems of Cats and Dogs in British Columbia, Canada." *Parasitology Research* 103(4): 889–98.
Latreille, P.A. 1812. "Mémoire sur un insecte que les Anciens réputaient fort vénimeux et qu'ils nommaient Bupreste." *Annales du Muséum d'Histoire Naturelle Paris* 19: 129–43.
Lattimore, R., ed. 1976. *The Iliad*. Chicago, IL: University of Chicago Press.
Laubin, R. and Laubin, G. 1980. *American Indian Archery*. University of Oklahoma Press.
Lauwerier, R.C.G.M. 1988. *Animals in Roman Times in the Dutch Eastern River Area*. Amersfoort: ROB.
Lawrence-Mathers, A. and Escobar-Vargas, C. 2014. *Magic and Medieval Society*. London: Routledge.
Lea, H.C. 1883. *Studies in Church History: The Rise of the Temporal Power, Benefit of Clergy, Excommunication, The Early Church and Slavery*. Philadelphia: Henry C. Lea's Son & Co. Available online: https://archive.org/details/cu31924029384975 (accessed November 1, 2018).
Leeson, P.T. 2013. "Vermin Trials." *Journal of Law and Economics* 56(3): 811–36.
Lefkowitz, J.B. 2014. "Aesop and Animal Fable." In G.L. Campbell (ed.), *The Oxford Handbook of Animals in Classical Thought and Life*, 1–23. Oxford: Oxford University Press.
Leick, G. 1998. *A Dictionary of Ancient Near Eastern Mythology*. London: Routledge.
Leneghan, R.T., ed. 1967. *Caxton's Aesop*. Cambridge, MA: Harvard University Press.
Liebert, R.S. 2010. "Apian Imagery and the Critique of Poetic Sweetness in Plato's Republic." *Transactions of the American Philological Association* 140(1): 97–115.
Lima, J.P. 2009. "La doctrina zoológica en la obra de San Alberto Magno." *STVDIVM, Revista de Humanidades* 15: 29–51.
Lindberg, D. 1992. *The Beginnings of Western Science*. Chicago, IL: University of Chicago Press.
Linow, W., ed. 1970. *Þe desputisoun bitwen þe bodi and þe soule*. Amsterdam: Rodopi.

Liu, X. 2001. "The Silk Road: Overland Trade and Cultural Interactions in Eurasia." In M. Adas (ed.), *Agricultural and Pastoral Societies in Ancient and Classical History*, 151–79. Philadelphia: Temple University Press.

Lucas, A.T. 1965. "Washing and Bathing in Ancient Ireland." *The Journal of the Royal Society of Antiquities of Ireland* 95: 65–114.

Malaxecheverría, I. 1999. *Bestiario Medieval*. Madrid: Ediciones Siruela.

Mann, J. 2009. *From Aesop to Reynard: Beast Literature in Medieval Britain*. Oxford: Oxford University Press.

Maraglino, V., ed. 2012. *La naturalis historia di Plinio nella tradizione medievale e umanustica*. Italy: Cacucci.

Marchenay, P. 1979. *L'Homme et l'abeille*. Paris: Berger-Levrault.

Mariño Ferro, X.R. 1996. *El simbolismo animal: creencias y significados en la cultura occidental*. Madrid: Encuentro.

Maspero, F., trans. 2011. *Plinio il Vecchio: Storie naturali (Libri VIII–XI). Testo latino a fronte*. BUR Biblioteca Universitaria Rizzoli.

Mayhoff, K.F.T., ed. 1906. *Naturalis Historia*. Leipzig: Teubner. Perseus Digital Library. Available online: http://www.perseus.tufts.edu/hopper/text?doc=Perseus%3Atext%3A1999.02.0138 (accessed March 5, 2018).

McCartney, E.S. 1920. "Spontaneous Generation and Kindred Notions in Antiquity." *Transactions of the American Philological Association* 51: 101–15.

McCormick, M. 2002. *Origins of the European Economy: Communications and Commerce AD 300–900*. Cambridge and Harvard: Cambridge University Press and Harvard Historical Studies.

McCulloch, F. 1962. *Mediaeval Latin and French Bestiaries*. Chapel Hill, NC: University of North Carolina Press.

McGavin, G.C. 2001. *Essential Entomology*. Oxford: Oxford University Press.

McGhee, R. 1984. "Contact between Native North Americans and the Medieval Norse: A Review of the Evidence." *American Antiquity* 49(1): 4–26.

McKitterick, R. 2008. *Charlemagne: The Formation of an European Identity*. Cambridge, MA: Cambridge University Press.

McLeod, M.P., Qin X., Karpathy, S.E., Gioia, J., Highlander, S.K., Fox, G.E., McNeill, T.Z., Jiang H., Muzny, D., Jacob, L.S., Hawes, A.C., Sodergren, E., Gill, R., Hume, J., Morgan, M., Fan G., Amin, A.G., Gibbs, R.A., Hong C., Yu X.J., Walker, D.H., and Weinstock, G.M. 2010. "Genome Sequence of *Rickettsia Typhi* and Comparison with Sequences of other Rickettsiae." *Journal of Bacteriology* 186(17): 5842–55.

Meaney, A. 1992. "The Anglo-Saxon View of the Causes of Illness." In S. Campbell et al. (eds.), *Health, Disease and Healing*, 12–25. Basingstoke: Macmillan.

Mellinkoff, R. 1985. "Demonic Winged Headgear." *Viator* 16: 367–81.

Mellinkoff, R. 2004. *Averting Demons: The Protective Power of Medieval Visual Motifs and Themes*. Los Angeles, CA: Wipf & Stock Publishers.

Mellor, P.A. and Shilling, C., eds. 1997. *Re-Forming the Body: Religion, Community and Modernity*. London and Delhi: Sage.

Meri, J.W. 2006. *Medieval Islamic Civilization: An Encyclopedia*, vol. 1. New York: Routledge.

Mermier, G.R. 1977. *A Medieval Book of Beasts: Pierre De Beauvais' Bestiary*. Lewiston, NY: Edwin Mellen Press.

Merrifield, M. 1849. *Original Treatises on the Art of Painting*. Mineola, NY: Dover Publications Inc.

Mezzalira, F. 2002. *Beasts and Bestiaries, The Representation of Animals from Prehistory to the Renaissance*. Torino: Umberto Allemandi.
Migne, J.-P. 1845. "S. Eusebii Hieronymi Stridonensis Presbyteri. Adversus Jovinianum. Libri Duo." In *Patrologia Latina*, 23. Available online: http://www.documentacatholicaomnia.eu/02m/0347-0420,_Hieronymus,_Adversus_Jovinianum_Libri_Duo,_MLT.pdf (accessed November 1, 2018).
Millward, J.A. 2013. *The Silk Road: A Very Short Introduction*. Oxford: Oxford University Press.
Mitchell, P.D. 2004. *Medicine in the Crusades: Warfare, Wounds, and the Medieval Surgeon*. Cambridge, MA: Cambridge University Press.
Mitchell, P.D. 2015. "Human Parasites in Medieval Europe: Lifestyle, Sanitation and Medical Treatment." *Advances in Parasitology* 90: 389–420.
Mittman, A.S. 2006. *Maps and Monsters in Medieval England*. Abingdon: Taylor & Francis.
Mode, H. 1975. *Fabulous Beasts and Demons*. London: Phaidon Press.
Monserrat, V.J. 2009a. "Los artrópodos en la Historia y en el Arte de la Ciudad de Venecia." *Boletín de la Sociedad Entomológica Aragonesa* 44: 603–28.
Monserrat, V.J. 2009b. "Los Artrópodos en la vida y en la obra de Vincent Van Gogh." *Boletín de la Sociedad Entomológica Aragonesa* 44: 629–42.
Monserrat, V.J. 2009c. "Los artrópodos en la obra de Hieronymus van Aken (El Bosco)." *Boletín de la Sociedad Entomológica Aragonesa* 45: 589–615.
Monserrat, V.J. 2010a. "Los neurópteros (Insecta: Neuroptera) en el arte." *Boletín de la Sociedad Entomológica Aragonesa* 46: 635–60.
Monserrat, V.J. 2010b. "Sobre los artrópodos en el tatuaje." *Boletín de la Sociedad Entomológica Aragonesa* 47: 477–97.
Monserrat, V.J. 2010c. "Los artrópodos en la Historia y en el Arte de la Ciudad de Florencia." *Boletín de la Sociedad Entomológica Aragonesa* 47: 499–549.
Monserrat, V.J. 2011a. "Sobre los artrópodos en los inicios de la abstracción y la figuración humana." *Boletín de la Sociedad Entomológica Aragonesa* 48: 1–45.
Monserrat, V.J. 2011b. "Sobre los artrópodos en la obra de Heródoto y su tiempo." *Boletín de la Sociedad Entomológica Aragonesa* 48: 525–43.
Monserrat, V.J. 2011c. "Sobre los artrópodos en la Arquitectura Ibérica." *Boletín de la Sociedad Entomológica Aragonesa* 49: 465–93.
Monserrat, V.J. 2012a. "Los artrópodos en la numismática de Grecia y Roma Clásicas." *Boletín de la Sociedad Entomológica Aragonesa* 50: 591–629.
Monserrat, V.J. 2012b. "Los artrópodos en la mitología, la ciencia y el arte de Mesopotamia." *Boletín de la Sociedad Entomológica Aragonesa* 51: 421–55.
Monserrat, V.J. 2013a. "Los artrópodos en la mitología, las creencias y el arte de los fenicios." *Boletín de la Sociedad Entomológica Aragonesa* 52: 347–71.
Monserrat, V.J. 2013b. "Los artrópodos en la mitología, las creencias, la ciencia y el arte del Antiguo Egipto." *Boletín de la Sociedad Entomológica Aragonesa* 52: 373–437.
Monserrat, V.J. 2013c. "Los artrópodos en la mitología, las creencias, la ciencia y el arte de los etruscos y la Roma antigua." *Boletín de la Sociedad Entomológica Aragonesa* 53: 363–412.
Monserrat, V.J. 2013d. "Sobre los artrópodos en la alfarería y la cerámica popular de la Península Ibérica." *Boletín de la Sociedad Entomológica Aragonesa* 53: 413–41.
Monserrat, V.J. 2014. "Los artopodos de los Beatos." *Boletín de la Sociedad Entomológica Aragonesa* 54: 469–503.

Monserrat, V.J. 2016. "Los artrópodos en los Libros Iluminados de la Edad Media europea." *Boletín de la Sociedad Entomológica Aragonesa* 58: 259–331.

Monserrat, V.J. 2016a. "Los artrópodos en los libros iluminados de la Edad Media europea." *Boletín de la Sociedad Entomológica Aragonesa* 58: 259–331.

Monserrat, V.J. 2016b. "Los artrópodos en la obra de Aristocles Podros (Platón)." *Boletín de la Sociedad Entomológica Aragonesa* 59: 321–49.

Monserrat, V.J. and Aguilar, J. 2007. "Sobre los artrópodos en el Grafiti Ibérico." *Boletín de la Sociedad Entomológica Aragonesa* 41: 497–509.

Monserrat, V.J. and Melic, A. 2012. "Las arañas en la cultura y el arte de Occidente (Chelicerata: Araneida)." *Boletín de la Sociedad Entomológica Aragonesa* 50: 631–73.

Montanari, M., ed. 1994. *The Culture of Food.* London: Wiley-Blackwell.

Montanari, M., ed. 1999. "Il messaggio tradito. Perfezione cristiana e rifiuto della carne*La Sacra Mensa: Condotte Alimentari e Pasti Rituali Nella Definizione Dell'Identità Religiosa*, 99–130. Modena: Fondazione Collegio S. Carlo – Banca Popolare dell'Emilia Romagna.

Moore, W. and Debboun, M. 2006. "History of Insect Repellents." In M. Debboun et al. (eds.), *Insect Repellents: Principles, Methods, and Uses*, 3–29. Boca Raton, FL: CRC Press.

Morge, G. 1973. "Entomology in the Western World in Antiquity and in Medieval Times." In R.F. Smith et al. (eds.), *History of Entomology*, 37–80. Palo Alto, CA: Annual Reviews Inc.

Mortimer, I. 2012. *The Time Traveller's Guide to Medieval England: A Handbook for Visitors to the Fourteenth Century*. London: Random House.

Moser, S. 1998. *Ancestral Images: The Iconography of Human Origins*. Stroud: Sutton Publishing Ltd.

Murga, P. 1983. *Símbolos*. Madrid: Rioduero.

Muthesius, A.M. 1992. "Silk, Power and Diplomacy in Byzanthium." *Textile Society of America Symposium Proceedings* 580: 99–110. Available online: http://digitalcommons.unl.edu/tsaconf/580 (accessed April 1, 2018).

Nazari, V. 2014. "Chasing Butterflies in Medieval Europe." *Journal of the Lepidopterists' Society* 68: 223–21.

Nelson, L.H., trans. 1996. *Herman of Tournai: The Restoration of the Monastery of Saint Martin of Tournai*. Washington, DC: Catholic University of America Press.

Neri, J. 2011. *The Insect and the Image: Visualizing Nature in Early Modern Europe, 1500–1700*. Minneapolis: University of Minnesota Press.

Neufeld, E. 1978. "Apiculture in Ancient Palestine." *Ugarit Forschungen* 10: 219–47.

Neville, J. 1999. *Representations of the Natural World in Old English Poetry*. Cambridge, MA: Cambridge University Press.

Newfield, T.P. 2015a. "Domesticates, Disease and Climate in Early Post-Classical Europe: The Cattle Plague of c.940 and its Environmental Context." *Post-Classical Archaeologies* 5: 95–126.

Newfield, T.P. 2015b. "A Cattle Panzootic in Early Fourteenth Century Europe." *Agricultural History Review* 57(2): 155–90.

Newfield, T.P. 2017. "Malaria and Malaria-like Disease in the Early Middle Ages." *Early Medieval Europe* 25(3): 251–300.

Nider, J. 1484. *Formicarius*. Ausburg: Antonius Sorg.

NMF. n.d. *Nordiska Museets, Folkminnessamling* (Archive of Ethnological manuscript notes). Stockholm.

O'Dowd, M.J. 2001. *The History of Medications for Woman, Materia Medica Woman.* New York: Parthenon Pub. group.

Olsen, P.E., Hames, C.S., Benenson, A.S., and Genovese, E.N. 1996. "The Thucydides Syndrome: Ebola Déjà Vu? (or Ebola Re-emergent?)." *Emerging Infectious Diseases* 2: 155–6.

Osten-Sacken, C.R. 1894. *On the oxen-born bees of the ancients (bugonia) and their relation to* Eristalis tenax, *a two-winged insect.* Heidelberg: J. Hoerning.

O'Sullivan, A., McCormick, F., Kerr, T., and Harney, L. 2014. *Early Medieval Ireland AD 400–1100: The Evidence from Archaeological Excavations.* Dublin: Royal Irish Academy.

Oxford English Dictionary. 2018. Oxford University Press. Available online: http://www.oed.com (accessed March 11, 2018).

Page, R.I. 1995. *Chronicles of the Vikings: Records, Memorials and Myths.* Toronto: University of Toronto Press.

Page, S. 2007. "Good Creation and Demonic Illusions. The Medieval Universe of Creatures." In B. Resl (ed.), *A Cultural History of Animals in the Medieval Age,* 27–57. Oxford and New York: Berg.

Panagiotakopulu, E. 2004. "Pharaonic Egypt and the Origins of Plague." *Journal of Biogeography* 31: 269–75.

Panofsky, E. 1939. *Studies In Iconology: Humanistic Themes in the Art of the Renaissance.* New York: Harper & Row.

Panofsky, E. 1955. *Meaning in the Visual Arts.* New York: Anchor Books.

Panofsky, E. 2007. *La arquitectura gótica y la escolástica.* Madrid: Siruela.

Papagrigorakis, M.J., Yapijakis, C., Synodinos, P.N., and Baziotopoulou-Valavani, E. 2006. "DNA Examination of Ancient Dental Pulp Incriminates Typhoid Fever as a Probable Cause of the Plague of Athens." *International Journal of Infectious Diseases* 10: 206–14.

Papagrigorakis, M.J., Yapijakis, C., and Synodinos, P.N. 2008. "Typhoid Fever Epidemic in Ancient Athens." In D. Raoult and M. Drancourt (eds.), *Paleomicrobiology: Past Human Infections,* 161–73. Heidelberg: Springer Science & Business Media.

Parker, J.H. Rivington, J.G.F., and Rivington, J., trans. 1844. *Morals on the Book of Job by St. Gregory The Great, Translated With Notes and Indices. In Three Volumes.* Available online: http://www.lectionarycentral.com/GregoryMoraliaIndex.html (accessed November 1, 2018).

Parkhill, J., Wren, B.W., Thomson, N.R., Titball, R.W., Holden, M.T., Prentice, M.B., Sebaihia, M., James, K.D., Churcher, C., Mungall, K.L., Baker, S., Basham, D., Bentley, S.D., Brooks, K., Cerdeño-Tárraga, A.M., Chillingworth, T., Cronin, A., Davies, R.M., Davis, P., Dougan, G., Feltwell, T., Hamlin, N., Holroyd, S., Jagels, K., Karlyshev, A.V., Leather, S., Moule, S., Oyston, P.C., Quail, M., Rutherford, K., Simmonds, M., Skelton, J., Stevens, K., Whitehead, S., and Barrell, B.G. 2001. "Genome Sequence of *Yersinia Pestis*, the Causative Agent of Plague." *Nature* 413: 523–6.

Pastoureau, M. 2006. *Una historia simbólica de la Edad Media occidental.* Buenos Aires: Katz.

Patlagean, E. and Rouche, M. 1991. *Historia de la vida privada, La Alta Edad Media, tomo 2.* Madrid: Taurus.

Payne, A. 1990. *Medieval Beasts.* London: The British Library Publishing Division.

Peck, R., ed. 1991. "The Storie of Asneth." *Heroic Women from the Old Testament in Middle English Verse*. TEAMS Middle English texts Series. Michigan: Medieval Institute Publications. Available online: http://d.lib.rochester.edu/teams/text/peck-heroic-women-storie-of-asneth (accessed March 1, 2018).

Pelauzy, M.A. and Català Roca, F. 1977. *Artesanía Popular española*. Barcelona: Blume.

Pigler, A. 1964. "La mouche peinte: un talisman." *Bulletin du Musée national hongrois des Beaux-Arts* 24: 47–64.

Platearius, M. n.d. *Book of Simple Medicines*. Available online: https://gallica.bnf.fr/ark:/12148/btv1b6000422n/f1.item (accessed November 1, 2018).

Platina, B. 1474. *De honesta voluptate et valetudine*. Rome: Udalricus (or Uldericus) Gallus.

Pohlit, A.M., Lopes, N.P., Gama, R.A., Tadel, W.P., and Andrade Neto, V.F. 2011. "Patent Literature on Mosquito Repellent Inventions which Contain Plant Essential Oils: A Review." *Planta Medica* 77: 598–617.

Pollock, M.S. 2005. *Figuring Animals: Essays on Animal Images in Art, Literature, Philosophy & Popular Culture*. London: Palgrave Macmillan.

Porter, R. 1999. *The Greatest Benefit to Mankind: A Medical History of Humanity*. New York: W.W. Norton & Company.

Preece, R. 1999. *Animals and Nature: Cultural Myths, Cultural Realities*. Vancouver: UBC Press.

Preston, C. 2006. *Bee*. London: Reaktion Books.

Procopius. 2007. *The Secret History*. Trans. G.A. Williamson. London: Penguin.

Purchas, S. 1614. *Purchas his pilgrimage, or Relations of the world and the religions observed in all ages and places discovered*. London. Available online: https://archive.org/details/purchashispilgri00purc/page/n5 (accessed November 1, 2018).

Race, W.H., ed. 1997. *Nemean Odes. Isthmian Odes. Fragments*. Cambridge, MA: Harvard University Press. Loeb Classical Library. Available online: https://www.loebclassics.com/view/pindar-fragments/1997/pb_LCL485.229.xml?rskey=OhH1GK&result=1 (accessed March 11, 2018).

Radford, U.M. 1949. "The Wax Images Found in Exeter Cathedral." *The Antiquaries Journal* 29(3–4): 164–8.

Ransome, H.M. 1937. *The Sacred Bee in Ancient Times and Folklore*. Boston, MA and New York: Houghton Mifflin Company, and London: George Allen and Unwin Ltd.

Ransome, H.M. 2004. *The Sacred Bee in Ancient Times and Folklore*. New York: Dover Books. Reprint of the George Allen and Unwin, London, 1937 edition.

Raoult, D. and Roux, V. 1997. "Rickettsioses as Paradigms of New or Emerging Infectious Diseases." *Clinical Microbiology Reviews*, October: 694–719.

Raoult, D., Aboudharam, G., Crubezy, E., Larrouy, G., Bertrand, L., and Drancourt, M. 2000. "Molecular Identification by 'Suicide PCR' of *Yersinia Pestis* as the Agent of Medieval Black Death." *Proceedings of the National Academy of Sciences of the United States of America* 97(23): 12800–3.

Ratcliffe, B.C. 2006. "Scarab Beetles in Human Culture." *Coleopterists' Society Monograph* 5: 85–101.

Rawcliffe, C. 1995. *Medicine and Society in Later Medieval England*. Stroud: Sutton Publishing Ltd.

Rawcliffe, C. 2013. *Urban Bodies: Communal Health in Late Medieval English Towns and Cities*. Woodbridge: Boydell Press.

Rayborn, T. 2014. *Against the Friars: Antifraternalism in Medieval France and England*. Jefferson, NC: Macfarland.
Réau, L. 1997a. *Iconografía del arte cristiano, Iconografía de los santos, de la A a la F*. Barcelona: Ediciones Serbal.
Réau, L. 1997b. *Iconografía del arte cristiano, Iconografía de los santos, de la G a la O*. Barcelona: Ediciones Serbal.
Réau, L. 1998. *Iconografía del arte cristiano, Iconografía de los santos. De la P a la Z, Repertorios*. Barcelona: Ediciones Serbal.
Réau, L. 1999. *Iconografía del arte cristiano, Iconografía de la Biblia, Nuevo Testamento, T.1*. Barcelona: Ediciones Serbal.
Réau, L. 2000. *Iconografía del arte cristiano. Introducción general*. Barcelona: Ediciones Serbal.
Regalado, N.F. 2013. "Shaping Courtliness in Richard de Fournival's Bestiaire D'Amours." In D. O'Sullivan and L. Shepard (eds.), *Shaping Courtliness in Medieval France*, 255–70. Cambridge: D.S. Brewer.
Reilly, E. 2003. "The Contribution of Insect Remains to an Understanding of the Environment of Viking Age and Medieval Dublin." In S. Duffy (ed.), *Medieval Dublin* IV, 40–63. Dublin: Four Courts Press.
Reilly, E. 2012. "Fair and Foul: Analysis of Sub-Fossil Insect Remains From Troitsky XI–XIII, Novgorod (1996–2002)." In M. Brisbane et al. (eds.), *The Archaeology of Medieval Novgorod in Context: Studies in Centre/Periphery Relations*, 265–82. Oxford: Oxbow Books.
Reilly, E. 2018. *Living Conditions in Early Medieval Europe: A Case Study from Viking Age Fishamble Street, Dublin*. Oxford: Archaeopress.
Rendell, A.W., trans. 1928. "Concerning the Spider." *Physiologus: A Metrical Bestiary of Twelve Chapters by Bishop Theobald*. London: John Edwards.
Resl, B. 2007a. *A Cultural History of Animals in the Medieval Age*. Oxford and New York: Berg.
Resl, B. 2007b. "Beyond the Ark. Animals in Medieval Art." In B. Resl (ed.), *A Cultural History of Animals in the Medieval Age*, 179–201. Oxford and New York: Berg.
Resnick, I.M. and Kitchell, K.F. Jr., trans. 2008. *Albert the Great Questions Concerning Aristotle's on Animals*, vol. 9, *The Fathers of the Church Medieval Continuation*. Washington, DC: Catholic University of America Press.
Revilla, F. 2003. *Diccionario de iconografía y simbología*. Madrid: Cátedra.
Riché, P. 1978. *Daily Life in the World of Charlemagne*. Philadelphia: University of Pennsylvania Press.
Roaf, M. 2000. *Atlas Cultural de Mesopotamia y el Antiguo Oriente Medio*. Barcelona: Optima.
Roberts, L.D. 1982. *Approaches to Nature in the Middle Ages*. New York: Medieval and Renaissance Texts and Studies.
Robinson, J.H., ed. 1969. *Petrarch: The First Modern Scholar and Man of Letters*. New York: Greenwood.
Robinson, M.W. 1961. *Fictitious Beasts: A Bibliography*. London: The Library Association.
Rooney, K. 2007. "Tradition and Innovation in the Middle English Debates of Mortality." In J. Harris and K. Cawsey (eds.), *Transmission and Transformation in the Middle Ages: Texts and Contexts*, 157–74. Dublin: Four Courts Press.
Ross, E.H. 1914. "House Flies and Disease." *Journal of the Royal Society of Arts* 62: 388–97.

Ross, J.L. 1971. *Pigment Used in Latest Medieval Western European Manuscript Illumination*. Austin, TX: University of Texas.

Rowland, B. 1974. *Animals with Human Faces: A Guide to Animal Symbolism*. London: Allen and Unwin.

Roy, W. 1994. *Signifying Animals: Human Meaning in the Natural World*. London: Routledge.

Ruskin, J. 2000. *Las piedras de Venecia*. Valencia: Consejo General de la Arquitectura Técnica de España.

Ryan, J.K., trans. 1960. *The Confessions of St. Augustine*. New York: Image Books.

Rytting, J.R. 2000. "A Disputacioun betwyx þe Body and Wormes: A Translation." *Comitatus: A Journal of Medieval and Renaissance Studies* 31(1): 217–32.

Sabatini, S. 1994. "Women, Medicine and Life in the Middle Ages (500–1500 AD)." *American Journal of Nephrology* 14: 391–8.

Salisbury, J.E. 1996. "Human Animals of Medieval Fables." In N.C. Flores (ed.), *Animals in the Middle Ages*, 49–66. London and New York: Routledge.

Sallares, R. and Gomzi, S. 2001. "Biomolecular Archaeology of Malaria." *Ancient Biomolecules* 3: 195–213.

Sallares, R. 2002. *Malaria and Rome: A History of Malaria in Ancient Italy*. Oxford: Oxford University Press.

Santiago-Álvarez, C. 2012. "Los insectos y otros artrópodos en el refranero español." *Boletín de la Sociedad Entomológica Aragonesa* 51: 377–90.

Sarris, P. and Williamson, G.A., eds. 2007. *Procopius: The Secret History*. New York: Penguin Classics.

Scanlan, J.J., trans. 1987. *Man and the Beasts: De animalibus, Books 22–26 (Medieval and Renaissance Texts and Studies, vol. 47)*. Center for Medieval and Early Renaissance Studies, State University of New York: Binghamton.

Scarborough, J. 1977. "Some Beetles in Pliny's *Natural History*." *The Coleopterists' Bulletin* 31(4): 293–6.

Scarborough, J. 1979. "On the History of Early Entomology, Chiefly Greek and Roman with a Preliminary Bibliography." *Melsheimer Entomological Series* 26: 17–27.

Schiavina, G. 1616. "Annales Alexandrini." In V.F. Ponzilioni (ed.), *Historiae Patriae Monumenta, XI, Scriptores IV*. Turin 1863.

Schimitschek, E. 1968. "Insekten als Nahrung, in Brauchtum, Kult and Kultur." *Handbuch der Zoologie* 4 (2)1/10: 1–62.

Schimitschek, E. 1974. "Mantis in Kult and Mythe der Buschmiinner." *Zeitschrift Für Angewandte Entomologie* 76: 337–47.

Schimitschek, E. 1980. "Manna." *Anzeiger für Schädlingskunde, Pflanzenschutz, Umweltschutz* 53: 113–21.

Schipper, W. 1997. *Annotated copies of Rabanus Maurus's De rerum naturis', in English Manuscript Studies 1100–1700*. London: British Library.

Schrader, J.L. 1986. "A Medieval Bestiary." *Metropolitan Museum Art Series: Metropolitan Museum of Art Bulletin* 44(1): 1–56.

Scott, S. and Duncan, C.J. 2001. *Biology of Plagues: Evidence from Historical Populations*. Cambridge, MA: Cambridge University Press.

Seeley, T. 2010. *Honeybee Democracy*. Princeton, NJ: Princeton University Press.

Segni, G.B. 1605. *Discorsi intorno alla carestia e fame*. Bologna: Heredi di Giovanni Rossi.

Seseña, N. 1997. *Cacharrería popular: la alfarería de basto en España*. Madrid: Alianza Editorial.

Seymour, M.C., ed. 1975. *On the Properties of Things: John Trevisa's Translation of Bartholomeaus Angelicus's De Proprietatibus Rerum*, vol. 1. Oxford: Clarendon Press.

Shailor, B. 2002. *The Medieval Book*. Toronto, Buffalo, London: University of Toronto Press.

Shank, M. 2015. "That There Was No Scientific Inquiry Between Greek Antiquity and the Scientific Revolution." In R.L. Numbers and K. Kampourakis (eds.), *Newton's Apple and Other Myths about Science*, 7–15. Cambridge, MA: Harvard University Press.

Shaver Crandell, A. 1989. *La Edad Media*. Barcelona: Gustavo Gili.

Shorey, P., ed. 1969. *Republic. In Plato in Twelve Volumes*, vol. 10. Cambridge, MA: Harvard University Press. Perseus Digital Library. Available online: http://www.perseus.tufts.edu/hopper/text?doc=Perseus:text:1999.01.0168 (accessed March 7, 2018).

Siganos, A. 1985. *Les mythologies de l'insecte, histoire d'une fascination*. Paris: Libraire des Méridiens.

Siraisi, N.G. 2012. "Medicine, 1450–1620, and the History of Science." *Isis* 103(3): 491–514.

Sisson, C.H., trans. 2008. *The Divine Comedy*. Oxford and New York: Oxford University Press.

Slavin, P. 2015. "Flogging a Dead Cow: Coping with Animal Panzootics on the Eve of the Black Death." In A. Brown et al. (eds.), *Coping with Crisis: Re-evaluating the Role of Crises in Economic and Social History*, 111–35. Martlesham: Boydell and Brewer.

Slavin, P. 2016. "Epizootic Landscapes: Sheep Scab and Regional Environment in England in 1279–1280." *Landscapes* 17(2): 156–70.

Smith, A.E. and Secoy, D.M. 1975. "Forerunners of Pesticides in Classical Greece and Rome." *Journal of Agricultural and Food Chemistry* 33(6): 1050–5.

Smith, A.J. 1996. *The Flea*. In *John Donne: The Complete English Poems*. Oxford: Oxford University Press.

Smith, D. 2012. *Insects in the City: An Archaeoentomological Perspective on London's Past*. British Archaeological Reports British Series 561. Oxford: Archaeopress.

Smith, D. 2013. "Defining an Indicator Package to Allow Identification of 'Cesspits' in the Archaeological Record." *Journal of Archaeological Science* 40: 526–43.

Smith, D. and Kenward, H.K. 2011. "Roman Grain Pests in Britain: Implications For Grain Supply and Agricultural Production." *Britannia* 42: 243–62. DOI: 10.1017/S0068113X11000031.

Smith, P.H. 2000. "Artists as Scientists: Nature and Realism in Early Modern Europe." *Endeavour* 24: 13–21.

Smith, R.F., Mittler, T.E., and Smith, C.N. 1973. *History of Entomology*. Palo Alto, CA: Annual Reviews.

Smith, V. 2007. *Clean: A History of Personal Hygiene and Purity*. Oxford: Oxford University Press.

Snell, M. 2017. "Silk Production and Trade in Medieval Times." Thought Co. Available online: https://www.thoughtco.com/silk-lustrous-fabric-1788616 (accessed November 1, 2018).

Somme, L. 2004. *Entomologiens Historie i Norge, Norsk Entomologisk Föreningen 1904–2004*. Oslo: Norsk Entomologisk Föreningen.

Sorabji, R. 1993. *Animal Minds and Human Morals: The Origins of the Western Debate*. Cornell Studies in Classical Philology. Ithaca, NY: Cornell University Press.

Soul and Body I, n.d. folio 101v, *Vercelli Book*. Available online: http://www.apocalyptic-theories.com/literature/soul1/soul1a.html (accessed March 5, 2018).

Spangenberg, L. 2010. "XPI autem generatio: The Book of Kells and the Chi-Rho Page." Digital Medievalist Blog, December 25. Available online: https://www.digitalmedievalist.com/2010/12/25/xpi-autem-generatio-the-book-of-kells-and-the-chi-rho-page/ (accessed November 1, 2018).

Speis, G. 2016. *Beekeeping on the Island of Andros: An Ethnographic Approach*. Kaïreios Library and the Eva Crane Trust.

Spiegel, H., ed. 1987. *Fables: Marie de France*. Toronto, Buffalo, and London: University of Toronto Press.

Sprecher, E. and Taroni, G. 2004. *Lucanus cervus depictus*. Como: G. Taroni editore.

Steel, C., Guldentops, G., and Beullens, P. 1999. *Aristotle's Animals in the Middle Ages and Renaissance*. Leuven: Leuven University Press.

Steel, K. 2013. "Abyss: Everything is Food." *Postmedieval: A Journal of Medieval Cultural Studies* 4: 93–104.

Strunz, F. 1925. *Albertus Magnus: Weisheit und Naturforschung im Mittelalter*. Wien: Karl König.

Swift, R.H. 1931. "The Sacred Beetles of Egypt." *Bulletin of the Southern California Academy of Sciences* 30: 1–14.

Szimai, J.A. 1999. *The Archaeology of Medieval Bookbinding*. New York: Routledge.

Szittya, P.R. 1986. *The Antifraternal Tradition in Medieval Literature*. Princeton, NJ: Princeton University Press.

Tacitus, C., Church, A.J., and Brodribb, W.J. 2008. *The Annals Book. The Reigns of Tiberius, Claudius and Nero by Cornelius Tacitus*. Oxford: Worlds Classics.

Taylor, A.L. 2013. *Epic Lives and Monasticism in the Middle Ages, 800–1050*. New York: Cambridge University Press.

Thompson, D.V. 1956. *The Materials and Techniques of Medieval Painting*. New York: Dover Publications.

Thompson, V. 2012. "The Gravestone, the Grave and the *Wyrm*." In *Death and Dying in Later Anglo-Saxon England*, pp. 132–68. Suffolk and New York: Boydell and Brewer.

Thorley, J. 1971. "The Silk Trade Between China and the Roman Empire at its Height, 'Circa' A.D. 90–130." *Greece & Rome* 18(1): 71–80.

Throop, P. 1998. *Hildegard von Bingen's Physica: The Complete Translation of Her Classical Work on Health and Healing*. Rochester, VT: Healing Arts Press.

Tillhagen, C.H. 1958. *Folklig Läkekonst*. Stockholm: Nordiska Museet, 12.

Tod, M.N. 1939. "The Scorpion in Graeco-Roman Egypt." *Journal of Egyptian Archaeology* 25: 55–61.

Toman, R. 2007. *Romanesque Architecture Sculpture Painting*. Potsdam: Ullmann and Könemann.

Torralba Burrial, A. 1998. "Los artrópodos en la Mitología Judeocristiana (Antiguo Testamento)." *Boletín de la Sociedad Entomológica Aragonesa* 23: 49–51.

Turpyn, R. 1541. *The chronicle of Calais, in the reigns of Henry VII and Henry VIII to the year 1540*. Available online: https://archive.org/details/chronicleofcalai00turprich/page/n3 (accessed November 1, 2018).

Twigg, G. 1985. *The Black Death: A Biological Reappraisal*. London: Schocken Books.

Van Huis, A., Cressman, K., and Magor, J.I. 2007. "Preventing Desert Locust Plagues: Optimizing Management Interventions." *Entomologia Experimentalis et Applicata* 122: 191–214.
Vatomsky, S. 2017. "When Societies Put Animals on Trial." *JSTOR daily*. Available online: https://daily.jstor.org/when-societies-put-animals-on-trial (accessed November 1, 2018).
Vedeler, M. 2014. *Silk for the Vikings (Ancient Textiles Series)*. Oxford: Oxbow Books.
Verdon, T. 2005. *María en el arte europeo*. Barcelona: Editorial Electa.
Vernon, F.G. 1979. "Beekeeping in 1269–1270 at Beaulieu Abbey in England." *Bee World* 60(4): 170–5.
Vorágine, J. De. 2014. *La Leyenda Dorada*. Madrid: Alianza Editorial.
Waddell, H.J. 1946. *Beast and Saints*. London: Constable and Co.
Wagner, D.M., Klunk, J., Harbeck, M., Devault, A., Waglechner, N., Sahl, J.W., Enk, J., Birdsell, D.N., Kuch, M., Lumibao, C., Poinar, D., Pearson, T., Fourment, M., Golding, B., Riehm, J.M., Earn, D.J., Dewitte, S., Rouillard, J.M., Grupe, G., Wiechmann, I., Bliska, J.B., Keim, P.S., Scholz, H.C., Holmes, E.C., and Poinar, H. 2014. "*Yersinia Pestis* and the Plague of Justinian 541–543 AD: A Genomic Analysis." *Lancet Infectious Diseases* 14(4): 319–26.
Walker Vadillo, M.A. 2013. "Apes in Mediaval Art." Medieval Animal Data Network (blog on Hypotheses.org), October 28. Available online: http://mad.hypotheses.org/172 (accessed November 1, 2018).
Wallis, F. 2010. *Medieval Medicine: A Reader (2010)*. Toronto: University of Toronto Press.
Walløe, L. 2008. "Medieval and Modern Bubonic Plague: Some Clinical Continuities." *Medical History*, Supp. 27: 59–73.
Walsh, J.J. 1920. *Medieval Medicine*. London: A. & C. Black Ltd.
Ward, J. 1902. *The Sacred Beetle: A Popular Treatise on Egyptian Scarabs in Art and History*. New York: Scribner. Available online: https://archive.org/details/sacredbeetlepopu00ward (accessed November 1, 2018).
Watson, W., Mitchell, T.C., and Cherry, J. 2015. Sigillography. *Encyclopedia Britannica*. Available online: https://www.britannica.com/topic/sigillography (accessed April 1, 2018).
Weisheipl, J.A. 1980. "Albertus Magnus and the Sciences. Commemorative Essays 1980." *Studies and Texts* 49: 103–27.
Weiss, H.B. 1912. "Some Ancient Beliefs Concerning Insects." *Bulletin of the Brooklyn Entomological Society* 8: 21–3.
Weiss, H.B. 1925. "The Bee, the Wasp, the Ant, Insects of the Physiologus." *Journal of the New York Entomological Society* 33(4): 238–42.
Weiss, H.B. 1927. "The Scarabaeus of the Ancient Egyptians." *American Naturalist* 61: 353–69.
Weiss, H.B. 1930. "Insects and Witchcraft." *Journal of the New York Entomological Society* 38(2): 127–33.
Werness, H.B. 2006. *The Continuum Encyclopedia of Animal Symbolism in Art*. New York: Continuum.
Wessèn, E. 1966. *Våra ord, deras utal och ursprung*. Stockholm: Norsteds.
Wheatley, E. 2000. *Mastering Aesop: Medieval Education, Chaucer, and his Followers*. Gainesville, FL: Florida University Press.
Whiting, B.J. and Whiting, H.W., eds. 1968. *Proverbs, Sentences, and Proverbial Phrases from English Writings, Mainly before 1500*. Cambridge, MA: Belknap Press.

Whiting, M.F., Whiting, A.S., Hastriter, W., and Dittmar, K. 2008. "A Molecular Phylogeny of Fleas (Insecta: Siphonaptera): Origins and Host Associations." *Cladistics* 24: 1–31.

Wigglesworth, V.B. 2017. "Insect." *Encyclopedia Britannica*. Available online: https://www.britannica.com/animal/insect (accessed March 11, 2018).

Winston, M.L. 2014. *Bee Times: Lessons From the Hive*. Cambridge MA, and London: Harvard University Press.

Wittkower, R. 1979. "Maravillas de Oriente: estudio sobre la historia de los monstruos." In R. Wittkower et al. (eds.), *Sobre la Arquitectura en la Edad del Humanismo*, 265–311. Barcelona: Gustavo Gili.

Woold, A. 1996. *Heraldic Art and Design*. Crayford: Shaw & Sons.

Wren, C.L. and Bolton, W.F., eds. 1988. *Beowulf, with the Finnesburg fragment*. Exeter: University of Exeter Press.

Wright, T. and Halliwell, J.O., eds. 1841. "Flen Flyss." In *Reliqiæ Antiquæ*, vol. I, 91–102. London: John Russell.

Wyatt, N. 1999. *Space and Time in the Religious Life of the Near East*. London: Bloomsbury Academic.

Wyndham, R.J. 1960. "Strange Superstitions about Bees and Honey." *Gleanings in Bee Culture* 88: 723–7.

Yale Library. 2013. *Iron Gall Ink*. Available online: https://travelingscriptorium.library.yale.edu/2013/03/21/iron-gall-ink/ (accessed November 1, 2018).

Yanoviak, E. 2013. "More than Marginal: Insects in the Hours of Mary of Burgundy." *Antennae* 26: 86–102.

Yapp, W.B. 1985. "A New Look at English Bestiaries." *Medieum Aevum* 54: 1–19.

Zabludoff, M. 2008. *Beetles*. New York: Marshall Cavendish.

Zadoks, J.C. 2013. *Crop Protection in Medieval Agriculture. Studies in Pre-Modern Organic Agriculture*. Leiden: Sidestone Press.

Zambon, F. 2010. *El alfabeto simbólico de los animales: Los bestiarios de la Edad Media*. Madrid: Siruela.

Zarco-Cuevas, J. 1997. "Pleito que se puso en la Abadia de Párreces para el exterminio de la langosta. Año de 1650." *Boletín de la Sociedad Entomológica Aragonesa* 20: 351–61.

Zarin, C. 2006. *Saints Among the Animals*. New York: Atheneum/Richard Jackson Books.

Zinsser, H. 1935. *Rats, Lice and History: A Chronicle of Pestilence and Plagues*. New York: Black Dog and Leventhal.

ACKNOWLEDGMENTS

Chapter 1

May R. Berenbaum is profoundly grateful to Emeritus Professor Richard Burkhardt of the University of Illinois Department of History for reviewing her chapter and expunging all howling historical errors before publication and providing context for medieval science, as well as to retired Professor Richard Leskosky, whose classical education and graduate work in linguistics helped enormously (ex gratia) in dealing with Latin spellings and translations.

Chapter 7

Rachel Murray is grateful to Claudio Cataldi, Gareth Griffith, Cathy Hume, Ad Putter, Jane Wright, and to members of the "Medieval-Religion" mailing list for directing her to a number of examples presented in her chapter.

Chapter 8

Julio Ferrer is deeply indebted to Dr. Daniela Panelid, Stockholm, Sweden, as well as Ludovic Chauwin and Catherine Paulus from the Musée de l'Oeuvre de Notre Dame, Strasbourg, France, for the generous authorization to print the photographs of the work *Les Amants trépassés*, conservé dans cette Musée.

INDEX